IDEAS AND APPLICATIONS TOWARD SAMPLE PREPARATION FOR FOOD AND BEVERAGE ANALYSIS

Edited by **Mark T. Stauffer**

Ideas and Applications Toward Sample Preparation for Food and Beverage Analysis
http://dx.doi.org/10.5772/65587
Edited by Mark T. Stauffer

Contributors

Tanja Butt, Dinesh Bilehal, Buket Er Demirhan, Burak Demirhan, Hayriye Eda Şatana Kara, Zhen Xie, Renata Raina-Fulton, Wagna Piler Santos, Rita Nano, Olga De Smidt, Shirleen Theisinger, Nnemeka Edith Ihegwuagu, Gabriel Malomo, Teodora Emilia Coldea, Elena Mudura, Carmen Socaciu

Published by InTech
Janeza Trdine 9, 51000 Rijeka, Croatia

Publishing Process Manager Martina Usljebrka
Technical Editor SPi Global
Cover InTech Design team

First published December, 2017
Printed in Croatia
Legal deposit, Croatia: National and University Library in Zagreb

Additional hard copies can be obtained from orders@intechopen.com

Ideas and Applications Toward Sample Preparation for Food and Beverage Analysis, Edited by Mark T. Stauffer
p. cm.
Print ISBN 978-953-51-3685-9
Online ISBN 978-953-51-3686-6

For EU product safety concerns:
IN TECH d.o.o., Prolaz Marije Krucifikse Kozulić 3, 51000 Rijeka, Croatia,
info@intechopen.com or visit our website at intechopen.com.

Meet the editor

Mark T. Stauffer was born in 1957. He graduated in Chemistry from the University of Pittsburgh in 1979, worked in industry for 12 years, and then returned to Pitt, receiving a PhD degree in Chemistry in 1998. He joined the chemistry faculty at Pitt-Greensburg in 2001, receiving tenure in 2007. Since 2001, he has collaborated on projects in archeology, foods, test kit evaluation, mine drainage, and data analysis. He and coauthors presented over 100 papers and posters at technical conferences and published 13 papers in peer-reviewed journals. He presents short courses on analytical data treatment at the annual Pittcon analytical chemistry conference. He conveys his enthusiasm for research and teaching via mentoring undergraduate research and through his courses in analytical chemistry.

Contents

Preface

Sample preparation is a vital component of any method used in analytical chemistry. Without the appropriate methods and techniques for releasing a sought-after analyte from the sample matrix that holds it bound, subsequent detection and quantitation of that analyte would be downright impossible or, at the very least, hellishly difficult, that is, unless the detection method does not require modification of the sample in any manner. Imagine trying to determine concentrations of manganese in tea by atomic absorption or emission spectrometry without extraction of the manganese into an acidic aqueous medium that solubilizes the manganese in its ionic forms. The tools and procedures used for manipulation of a sample toward extraction of component analytes make possible the extraction of analytes from the matrix and, subsequently, the quantitative determination of those analytes. Of utmost importance, the sample preparation methods and techniques must be ones that yield the most accurate and precise results possible.

Sample preparation methods and techniques have become extremely diverse over the past century or so, covering a wide range of sample types and matrices and employing a huge range of laboratory techniques including, but by no means limited to, acid and base digestion methods, microwave-induced decompositions, and extractions utilizing a wide array of solvents and solutions. This short description is merely the tip of an immense "iceberg" of sample preparation methods and techniques employed for a plethora of inorganic, organic, biological, and environmental sample types, to name only a few. The myriad of foodstuffs and beverages available to the world at large constitutes a significant portion of the sample types for which exist a large number of sample preparation methodologies and into which considerable research effort is currently being invested.

The focus of this book is on methods and techniques involving preparation of samples for subsequent identification and/or quantitation of analytes and the applications of these sample preparation methods and techniques to a variety of research questions and issues. A selection of ideas and applications toward sample preparation methodologies will be presented in a series of original research and review chapters. Advances in various types of sample grinding technologies, e.g., cryogenic grinding, are presented, with a focus on reduction of sample size and efficient, effective sample homogenization to obtain the most accurate and precise analytical results possible. The current emphasis on "environmentally friendly" approaches to a plethora of problems and questions in research overall, and to chemical analysis in particular, is becoming widely utilized in development and application of sample preparation methodologies as well. The advantages and disadvantages of such "green" sample preparation schemes are reviewed as potential alternatives to more traditional sample preparation methods and techniques. The impact of bioaerosols on all aspects of the food and beverage industries, and the importance of, and immense need for, standardized protocols for bioaerosol monitoring as a result, is also presented and discussed in this book, along with a review of the state of sample preparation issues and research in the area of animal feeds. Applications of sample preparation methods and techniques, e.g., QuEChERS, toward determination of pesticide residues in chilies and other biological samples by chromatographic-mass spectrometric detection are presented in this book, as well as applications of microwave digestion and other sample preparation methodologies to studies of bioavailability of elements from various legumes, and a review of sample preparation schemes and detection methods for authenticity and quality testing of distilled alcoholic beverages.

The goal of this book is to present an overview of ideas and applications involving sample preparation of foods and beverages for subsequent chemical analysis. This text is a compilation of selected research articles and reviews covering current efforts in research on, and applications of, sample preparation methodologies for food and beverage analysis. The chapters in this book are divided into two broad sections. Section 1 deals with some ideas for methods and techniques that are applicable to problems that impact the analysis of foods and beverages and the food and beverage industries overall. Section 2 provides applications of sample preparation methods and techniques toward determination of specific analytes or classes of analytes in various foods and beverages. Overall, this book should serve as a source of scientific information for anyone involved in any aspect of analysis of foods and beverages.

I am most appreciative to Ms. Martina Usljebrka, the current Publishing Process Manager, and her predecessors, Ms. Mirena Čalmić and Ms. Ana Pantar, who supervised and organized publishing of all materials; assisted me and the authors in completion of our work in an easy, timely manner; and provided helpful advice and guidance throughout this project. I thank the authors for their wonderful contributions to this book and their prodigious efforts toward those contributions. I also wish to thank the technical editors who prepared these manuscripts for publication by InTechOpen Access Publisher. I thank my wife, Resa, also an analytical chemist, for her advice and support and my colleagues, support staff, and administrators at the University of Pittsburgh at Greensburg for their support throughout this endeavor. Last, but by no means least, I extend my thanks to a group of former colleagues from my industrial days in the 1980s, who taught me more about sample preparation and its paramount importance in chemical analysis than I ever thought I could learn. Thank you again, after all these many years!

Mark T. Stauffer, PhD
University of Pittsburgh at Greensburg,
Greensburg, Pennsylvania,
United States of America

Section 1

Ideas Toward Sample Preparation in Food and Beverage Analysis

Chapter 1

Overview of Green Sample Preparation Techniques in Food Analysis

Burak Demirhan, Hayriye Eda Şatana Kara and
Buket Er Demirhan

Additional information is available at the end of the chapter

http://dx.doi.org/10.5772/intechopen.68787

Abstract

Nowadays, the significance of food analysis could be emphasized in consequence of growing world population besides the increased consumer demands for the safe food. The reliability and accuracy of analysis are highly affected by sample preparation, extraction, enrichment, and isolation of the analytes. Traditional sample preparation techniques are not only costly but also time-consuming and generally labor-intensive, and furthermore, these techniques required high solvent content, which generates waste, pollutes sample, and enriches the analyte for the food analysis. In recent years, new extraction techniques have been discovered as an alternative to the conventional sampling procedure. Simple, fast, cost-effective and green (environmentally friendly) techniques can be preferred gradually instead of traditional methodologies in order to the extraction of the sample. The aim of the chapter will be to compile and discuss the advantages, pro and cons, and use of some sample preparation techniques that are relevant to the green chemistry.

Keywords: food, green chemistry, microextraction, microwave, ultrasound

1. Introduction

Analytical control analysis for food safety and quality is developing steadily. Sample preparation is one of the main steps of food analysis. Direct analysis of several compounds in foodstuff is very difficult without any sample preparation methods. Generally, a huge amount of toxic organic solvents is required in traditional techniques. Thus, these techniques are both costly and environmentally harmful, and produce waste [1]. The necessity of the novel analytical techniques in the food science is related to the demand of information about the process, quality control, adulteration, contamination, and food regulations. For this purpose, chemists, regulatory agencies, and quality control laboratories request faster, more powerful, clean, and inexpensive analytical procedure to meet this demand. Improvement in the modern analytical techniques causes clear development in the quality of analysis [2].

Green chemistry could be applied to the chemistry to decrease or eliminate the harmful substances in the chemical products design and process [3]. Green analytical chemistry arose from green chemistry in 2000. The key goals of greening analytical methods should be ensured by reductions of sample number, reagents, energy, waste, risk, and hazard [4]. Green extraction techniques are the alternative to classical sample preparation techniques [5]. These techniques provide opportunities to reduce or eliminate the chemical solvent usage while improving the quality of extract, efficiency of the methods, and extraction of products [6].

2. Green analytical chemistry and food analysis

Chemistry has long been perceived as dangerous, and chemical and toxic words are often associated with chemistry by humans. Several security precautions such as protecting clothing could be taken to reduce the risk. Therefore, the purpose of the green chemistry is minimizing the risk occurred during the chemical life cycle. Also, risk should be defined as the ability to create an adverse effect on human and environment [7]. Green chemistry can be determined as a design of chemical products and process to reduce or eliminate the formation and use of the harmful substances. This definition and green chemistry concept were first introduced in the early 1990s [7, 8]. Green chemistry is the methodology and chemical techniques to eliminate or reduce the use of a solvent, reagent, products, and by-products that harmful to the human health and environment. In brief, green chemistry is the use of the chemistry in order to avoid pollution. Analytical laboratories previously developed the green chemistry ideas and same philosophy. Environmental side effects of analytical methods are reduced by three ways: (i) to reduce the amount of solvent in sample pretreatment; (ii) to reduce the amount and toxicity of solvent and reagents during the measurement; and (iii) to develop alternative analytical procedures that avoid the use of solvent and reagents [9]. For this purpose, green chemistry has a set of principles to reduce or prevent the harmful substances which are used in the design, production, and application process of the chemical products [10]. Twelve principles of green chemistry were introduced in 1998 by Anastas and Warner [11] and they were given in **Figure 1** [7, 8, 11, 12]. Principles of the green analytical chemistry for sample preparation and final analysis stage are given in **Figure 2** [13].

Chemical substance reduction or elimination, effective energy consumption and waste management, and enhancement of the safety are key goals of the green analytical methods [4]. Food analysis has been serious for the purpose of quality control of raw and processed foods, specifying of the nutritional value of foods, and monitoring the food additives and toxic contaminants [14]. Life quality should be improved in developing countries due to the application of cheap, fast and environmental safety procedures during the analysis of foods [5]. Application of these principles should be actualized whole analytical process steps: sampling,

1. **Prevention-** Prevention of waste.
2. **Atom economy-** Designing synthetic method.
3. **Less hazardouschemical synthesis-** Synthesis less harmful chemical.
4. **Desiginig safer chemical-** Design of safer chemicals.
5. **Safer solvents and auxiliaries-** Use of safer solvents and auxiliaries such as separating agents.
6. **Design for energy efficiency-** Minimizing the energy requirements of chemical process.
7. **Use of renewable feedstocks-** Use of renewable materials.
8. **Reduce derivatives-** Minimizing the derivatization process.
9. **Catalysis-** Use of catalytic reagents.
10. **Design for degradation-** Designing innocuous degradation products.
11. **Real-time analysis for pollution prevention-** Real time analysis for in-process monitoring.
12. **Inherently safer chemistry for accident prevention-** Safer chemistry to prevent accidents

Figure 1. Twelve principles of green chemistry [7, 8, 11, 12].

Figure 2. Principles of the green analytical chemistry for sample preparation and final analysis stage [13].

preparing samples, separation, detection, and data analysis [15]. The purpose of the sample preparation is to enable to the isolation of the target analytes, to minimize the complexity of the samples and to prevent most of the matrix interferents, before the analytical detection (e.g., chromatographic techniques) [16]. Sample preparation has been considered as a time-consuming process of the analytical procedures among these steps. Therefore, simple and environmentally friendly techniques can be preferred gradually instead of traditional methodologies in order to extraction and preconcentration of analytes [17]. Separation techniques are applied for food analysis after sample preparation. Application of the principles of green chemistry into gas chromatography (GC) can be performed in many ways. Liquid chromatography (LC) is generally recognized less green than GC, due to the solvents requirement for the separation. On the other hand, LC offers more possibilities for "greening" [13]. Capillary zone electrophoresis (CZE) has also some advantages such as environmental friendliness, analysis time, and cost-effectiveness [18].

3. Sample preparation and greener approach

In general, foods cannot be analyzed without any presample preparation steps, because of diluted analytes and complex matrix structure of the foods [1]. Sample preparation is an extraction process, which extracts the chemical residues from the sample. Therefore, isolation of target residues and removal of interferents are ensured by sample preparation. Sample preparation has been the major part of the analytical procedures as well as separation and detection techniques, and effective sample preparation gives rise to reliable results and provides the instrumental performance [19]. The reliability and accuracy of analysis are highly affected by sample preparation, extraction, enrichment and isolation of the analytes [1]. Sample preparation should be recognized as possibly causative steps to problems and complications due to the time consumption, cost, contamination, and low extraction efficiency [16]. Several analytical steps such as purification, gel permeation chromatography (GPC), sulfuric acid treatment, and adsorption chromatography (alumina, silica gel, Florisil) should be applied as single or in combination with in order to avoid interferent compounds (e.g., lipids, carbohydrates, water, and chlorophyll), which can be extracted as well as the target analyte during the extraction of food samples. Acid digestion or saponification is a destructive method in order to remove the lipids [20].

Occasionally, transfer to the liquid phase should be necessary, by reason of difficult analysis of solid samples. Leaching the analyte (i.e., solid-liquid extraction or lixiviation) is one of the simpler, most widely applied sample treatments. Soxhlet extraction, which is basic reference against new leach methods, can be still employed in the routine analysis due to the lower costs and robustness. Soxhlet system is basic, easy to use and provides to use a large amount of sample. Unfortunately, there are several disadvantages such as long extraction periods and high solvent consumption [5]. Several extraction methods have been applied for the sample preparation in food analysis. Soxhlet and pressurized extraction techniques (e.g., supercritical fluid extraction (SFE) and accelerated solvent extraction (ASE)) have been employed to sample analysis. Large bore open-tubular glass liquid chromatographic (LC) columns operating at gravity pressure and liquid-liquid extraction (LLE) techniques are used in the liquid samples [16]. Conventional LLE and solid phase extraction (SPE) have been used widely in food analysis in order to prepare food samples [21].

In food analysis, traditional sample preparation techniques are not only costly, but also time-consuming and generally labor intensive, and furthermore, these techniques required high solvent content, which generates waste, pollutes sample, and enriches the analyte. Additionally, more than one clean-up steps are necessary before detection steps. Extraction of organic analytes from food samples generally begins with homogenization, and subsequently, exhaustive liquid extraction steps with one or more clean-up steps and purification of the extract before the detection of the analyte are required [20]. Consequently; several alternative techniques were developed to solve this trouble [2].

In recent years, some new extraction methods were discovered as an alternative to the conventional sampling procedure. Green extraction, which could be an alternative renewable sample preparation technique, improves the sensitivity and selectivity of analytical methods according to the classical sample preparation methods [5]. Current trends in the sample preparation have been focused on the low-cost operations owing to the miniaturization, automation, high efficiency performance, online analytical instruments, and extremely low- or nonsolvent consumption. Minimizing the sample preparation steps can be effective due to the reducing errors, time, and cost, and has some advantages in order to measure trace and ultra-trace analytes in complex matrices. Due to the disadvantages such as time-consuming procedures and excess use of liquid organic solvents, new sample preparation methods allow using less organic solvents and can be alternative to the conventional methods [21]. Microextraction methods (e.g., solid phase microextraction (SPME), stir bar sorptive extraction (SBSE), and liquid phase microextraction (LPME)) have been becoming important in order to prepare samples in comparison with conventional techniques [1]. By definition microextraction means, all modes of these techniques require using small volumes of extraction medium during the extraction conditions. The large number of variable parameters such as extraction time, temperature, pH, salt concentration, stirring rate, sample volume, etc. and relationships between them often require the avoiding from the classical approach in order to optimization, which does not take into account the interactions between the variables [22]. These methods are carried out even though the complexity of the samples of food analysis [1].

4. Sample preparation techniques relevant to green chemistry

4.1. Solid phase microextraction

Most common techniques for purification have been adsorption chromatography that uses solid phase extraction (SPE) techniques [20]. SPME methods should be recognized as a green approach for the sample preparation due to the reduction of the solvent consumption and waste production. Additionally, SPME has some advantages such as time-saving and cost-effective against conventional methods [17]. SPME can allow to extraction and enrichment process in a single step that produces solvent-free sample preparation. SPME was first developed in 1990 by Arthur and Pawliszyn [23]. Appropriate adsorbent phase coated silica fiber has been used in this technique. The analyte is directly extracted from the sample and then concentrated to the fiber coating [1, 24]. The efficiency of the analyte preconcentration can be depended on the several parameters such as the type of fiber, sample stirring, extraction time etc. in the SPME technique [25]. SPME should be defined as a sample preparation technique that uses fused silica fiber coated with appropriate stable phase. Volatile and semivolatile compounds in the food samples can be extracted by SPME in combination with GC and GC/mass spectrometry (MS). SPME is also used in combination with high-performance liquid chromatography (HPLC) and LC-MS to extract weak volatile or thermally unstable compounds which are unsuitable to the GC or GC-MS. SPME-HPLC interface has special desorption chamber, and it is used in order to desorption of the solvent before the HPLC analysis instead of thermal desorption in the injection port of the GC. In-tube SPME, which is new SPME-HPLC systems and is appropriate to the automation, has been recently developed using an open tubular fused silica capillary column. Automatic sample processing procedure not only reduces the total analysis duration but also provides more accuracy and precision according to the manual techniques [24].

Simplicity, speed, solvent free, high sensitivity, and small sample volume are the main advantages of the SPME [1, 26]. SPME techniques have also been significant due to the reproducibility, repeatability, and possible quantitative determination. Polar and nonpolar compounds in gaseous, liquid, and solid samples can be analyzed by SPME which is successfully combined with various analytical instruments such as GC, HPLC, etc [1].

SPME should be applied for various food analyses. SPME has lowest detection limits when compared to several common sample preparation methods which are important to the analysis of the main flavor and odor compounds [27].

There are two available extraction types in the fiber SPME; headspace extraction SPME (HS-SPME) is the first, and direct extraction SPME (DI-SPME) is the second [14, 24]. Furthermore, HS-SPME exposes a lower background than DI-SPME and is appropriate to the molecular extraction of more volatile analytes in the gas, liquid, and solid food samples. DI-SPME should be used to the extraction of semi or less volatile analytes in the liquid samples [24].

4.1.1. Direct extraction SPME

Fiber should be placed directly to the liquid samples in the direct extraction (DI-SPME). Agitation can be necessary to accelerate the extraction in the sample matrix [1]. Additionally, the natural flow of air should be sufficient to ensure balance to the volatile compounds in gaseous samples [1, 24].

4.1.2. Headspace extraction SPME

Headspace extraction (HS-SPME) is comprised of two equilibriums; the first equilibrium is between sample matrix and gaseous phase (headspace) above it, and the second equilibrium is between headspace and coating on the extracting fiber [1]. HS-SPME methods, in which 100 μm polydimethylsiloxane (PDMS) fiber is used, can be used in combination with GC or GC-MS for the purpose of the analysis of various foods [14]. A lifetime of the fiber can be extended in the HS-SPME technique, in consequence of not contacting with the sample directly. Less volatile compounds are also extracted directly from the solution in DI-SPME [25]. SPME should be used almost completely as headspace extraction in the food matrices (the presence of sugars, proteins, colorants, and other nonvolatiles), because of the specificity of this food matrix [1].

4.1.3. *In-tube SPME*

This is the effective sample preparation technique that uses the open tubular fused-silica capillary column as an extraction device [28]. Although in-tube SPME can be appropriated to these compounds, particles must be removed from samples by filtration before the extraction in order to avoid plugging of the flow line and capillary column during the extraction so that in-tube SPME has preferred to the extraction of the clean samples [14]. Tube design of the SPME can be used with very similar arrangements of the SPE; however, the main difference is to add a volume of the extraction phase due to the fact that the purpose of the SPME is never detailed extraction [29]. In-tube SPME is an ideal sample preparation technique in food analysis due to the fast operation and easy automation of technique, and being solvent free and inexpensive [28]. In-tube SPME, which is an appropriate technique for automation, includes automated sample handling to ensure reduced total analysis time and better accuracy and precision compared to manual techniques [14]. This can be facilitated by the design of this system [29]. In-tube SPME has two extraction modes, which are static and dynamic modes. In the static mode, mere diffusion is responsible for the transfer of the analytes to the stationary phase. The dynamic mode includes repeated draw/eject cycles for the purpose of sample extraction [30].

4.2. Stir bar sorptive extraction

Stir bar sorptive extraction (SBSE), which is the microextraction methods introduced by Baltussen et al. [31], fulfills the requirements of the green chemistry by removing excessive solvent usage, and reducing labor-intensive and time-consuming sample preparation steps. PDMS coated 10–40-mm-long magnetic stirrers can be used as mobile sorptive elements in the original SBSE. Therefore, the total volume of the sorbent is used to the extraction of the analyte [31]. SBSE is very easy to handle samples and allows great selectivity and sensitivity for complex matrices [32].

This sorptive extraction technique has been basically the same principle with SPME, but extraction capacity of the SBSE is higher [1]. In contrast to the coated fiber SPME, magnetic stir bar should be used in order to capture the analytes during the stirring in SBSE. Coated phase is usually PDMS fiber which has 50–250 times higher extraction volume compared to the SPME fiber, resulting in higher recoveries and higher sample loading capacity. Normally, SBSE can be applied to the extraction of semi-volatile and volatile organic compounds in the aqueous matrix of the foods. Because of this goal, stir bar can be basically added and rotated for the sample extraction, after a while molecules captured by bars should be thermally desorbed in the GC or added to the solvent for the LC. Manually operation in most cases is the main disadvantage of the SBSE [33]. Coated stir bar should be added to the sample for the purpose of the stirring and extraction (Direct SBSE) or exposed to the sample headspace (HS-SBSE) in SBSE technique [1]. The efficiency of the SBSE compared to the other sorptive techniques was investigated. Different types of organic compounds in aqueous solutions can be extracted by SBSE technique. The detection limit of SBSE may be reduced by its use in combination with thermos desorption-GC-MS [25]. SBSE is not popular technique as well as SPME, but it has been recognized as a green alternative technique for the extraction of pesticide residue in the sugarcane [9].

4.3. Liquid phase microextraction techniques

Liquid phase microextraction (LPME) is the alternative extraction techniques to the SPME and should be classified as three types: single drop phase microextraction (SDME), hollow fiber liquid phase microextraction (HF-LPME), and dispersive liquid-liquid microextraction (DLLME) [1]. All of the types of LPME, especially SDME, use organic solvents as microliter volume, resulting in being environmentally friendly [21]. LPME which was developed in 1996, is an easy, fast, efficient, and cheap sample preparation technique. Extraction, concentration and sample input can be integrated into a single step [34]. LPME term should be used as a little solvent volume of LLE (acceptor phase-water immiscible) to the extraction of the analytes from liquid solution (donor phase) [1]. Extraction in the LPME normally consists of between small amount of water-immiscible solvent and an aqueous phase containing target analyte. The acceptor phase is not only immersed for the direct extraction but also suspended on the sample for the headspace extraction. Receiving phase volume is ranged microliter or below; considering this, higher enrichment factors should be obtained due to the ratio of the high volume of sample to the acceptor phase [21]. Since then, different LPME approaches have been developed to analysis various compounds in foods: SDME, HF-LPME, DLLME, with each group having a variety of modifications [1, 35].

LPME advantages can be summarized as simple and highly selective extraction method; it has been combined with HPLC, capillary gas chromatography (GC), and capillary electrophoresis (CE), environmentally friendly due to less solvent usage, in which μL solvent is used to extraction of an analyte from various samples [36].

4.3.1. Single-drop microextraction

Single-drop microextraction (SDME) technique was developed in 1996 by Liu and Dasgupta, and this technique uses suspension of a microdrop (~1.3 μL) of water-immiscible organic solvent in an aqueous solution [37]. SDME has been a first successful application in the LPME technique in order to concentrate and purified of the analytes during the liquid and gaseous samples analysis [38]. SDME is a new, simple, fast, and environmentally friendly method, and effects of nature of organic solvents, microdrop volume, microdrop depth in the samples, extraction time, and stirring speed on the extraction efficiency have been separately demonstrated by Li et al. [39]. This method is successfully applied to the GC-MS for the purpose of the determining the phthalate esters in food samples [39].

Fiber should be necessary to the extraction of the analytes in both SPME and HF-LPME techniques, but only single microdrop is used as solvent acceptor phase in SDME, which is more simple, practicable and almost costless in comparison with SPME and HF-LPME [1]. SDME uses organic solvent at the end of the microsyringe and is developed from LPME technique. SDME should be classified as direct immersion SDME (DI-SDME) and headspace SDME (HS-SDME). Water immiscible solvent drop can be suspended apex of the microsyringe needle which is immersed in the aqueous sample in DI-SDME. Sample headspace or flowing air sample stream involves a microdrop of an appropriate solvent to extract the volatile compounds in the HS-SDME. Advantages of the HS-SDME are to choose a wide variety of solvents. On the contrary, the necessity of the different apparatus to extraction and injection is the disadvantages of the HS-SDME techniques [36].

4.3.2. *Dispersive liquid-liquid microextraction*

Recent years, dispersive liquid-liquid microextraction (DLLME) is a new miniaturize extraction technique which has been introduced by Rezaee and co-workers in 2006 [40]. DLLME basically depends on the three-component solvent systems (aqueous sample, dispersive solvent, and extractive solvent). The appropriate mixture of the extraction solvent (organic) and dispersive solvent (water-organic miscible solvent) can be injected into the aqueous sample, and thus, cloudy solvent should be formed. Subsequently, via the centrifuge, analytes are separated from the organic phase. In the extractive solvent, concentrated analytes should be injected to the GC, LC or electromigration instruments for the purpose of separation and detection [21].

4.3.3. *Hollow fiber liquid phase microextraction*

Hollow fiber liquid phase microextraction (HF-LPME) was introduced by Bjergaard and Rasmussen in 1999 [41]. The main basis of the hollow fiber-based LPME is to fill the little sample vial with the targeted liquid sample, and porous hollow fiber is placed into the samples. The volume of liquid sample and length of fiber are varying between 0.1 and 4 mL and 1.5–10 cm, respectively. Before the extraction, immersed a part of the hollow fiber to the organic solvent immobilizes the solvent on the hollow fiber and then excess of the solvent is removed [42]. HF-LPME is a simple and cheap technique that ensures analyte extraction from complex samples. The analyte can be extracted by extractant from liquid samples in the two phase LPME sampling mode. This extractant is into the porous hollow fiber which is made from polypropylene material supported with microinjector. In this sampling mode, acceptor phase is organic, so that this system is compatible with GC and HPLC in order to total analysis [43]. HF-LPME can be done in two modes such as static or dynamic, and the second one gives less operating time, while it cannot be automated, and therefore, it is necessary to optimize and control [44]. This technique is successfully used in the complex matrix such as foodstuffs for the purpose of the cleaning and extracting of the samples [1].

4.4. Ultrasound-assisted extraction

Over the last decade, application of ultrasound for extraction has increased, due to a number of disadvantages associated with conventional or other newer techniques, such as high capital investment and energy consumption, and the use of toxic organic substances used for extraction. In the preapplication steps, ultrasound-assisted extraction (UAE) is a method that ultrasound technique is applied, and this technique can be preferred in terms of being environmentally friendly and clean extraction [45]. Consequently, use of UAE has been recognized as green and economically viable alternative to conventional techniques in food [46]. Therefore, ultrasound is an easy to use, the multi-directional, flexible and low investment required technique when compared to the other extraction techniques such as SFE, PLE or ASE. Ultrasonic area of the spectrum is important because of the conventional applications. Ultrasound generally should be classified as low intensity sonication (<1 W/cm^2) and high density sonication (10–100 W/cm^2). High intensity sonication is performed to the extraction and process applications, while low intensity sonication is used as a nondestructive analytic technique for the quality assurance and process control. Ultrasound application enlarges the solvent selection range of generally recognized as safe (GRAS) instead of toxic organic solvents [45].

4.5. Microwave-assisted extraction

Microwave-assisted extraction (MAE) can be applied in order to the extraction of organic compounds from a different type of matrix. In this method, lower extraction time and the lower organic solvent are used compared to the conventional extraction [9]. Nowadays, MAE should be applied to solid samples as a versatile extraction technique and desorb analytes using electromagnetic radiation. While the microwave frequency varies between 300 MHz and 100 GHz that can be used the whole of the frequency, conventional ovens should only operate at 2.45 GHz. Very fast heating, high temperatures, and ease of operation are the main advantages of the MAE, and the only disadvantage of the MAE is the limited heating of the sample solvent due to the dielectric constant [47].

4.6. Supercritical fluid extraction

Supercritical fluid extraction (SFE) is green, easily and totally automatable analytical method [9]. Environmentally friendly sample preparation method (e.g., typically SFE) is a method that uses environmentally friendly solvents such as water. SFE can relatively eliminate the risk of activity loss using short extraction time, lower pressure and temperature, and it can protect the integrity of functional compounds of food [33]. SFE advantages are getting clean extract due to the reduced solvent usage and extraction time. There are no further clean-up steps in the extraction of the analytes. In this technique, nontoxic and nonpolluting extraction fluids such as carbon dioxide can be most widely used in the sample pretreatment [2, 20]. The main advantages of the SFE are quantitative, simple, fast, selective and environmentally friendly. SFE is used to the extraction of the pesticide residue from fruits [9].

4.7. Pressurized fluid extraction

Pressurized fluid extraction (PFE) is similar with soxhlet extraction except that the usage of the solvent is near the supercritical area. PFE could cause to higher extraction efficiency due to the lower solvent volume as 15–40 mL and short extraction time as 15–20 min. PFE is also known as ASE was first introduced in 1996. Additionally, ASE, known as pressurized solvent extraction (PSE), pressurized liquid extraction (PLE) and solvolytic extraction, is a solid-liquid extraction process which is operated at high temperatures (50–200°C) and pressure (10–15 MPa). While organic solvent is generally used in the ASE, pressurized hot water can be used in this technique. Main advantages of ASE technique are reducing the extraction time and solvent usage in comparison with the traditional extraction methods [9].

4.8. Cloud-point extraction

The cloud-point extraction (CPE) is other greener sample pretreatments, and it was first developed by Watanabe and Tanaka for the preconcentration of metal ions from aqueous samples [9].

CPE consists of three steps: (i) solubilize the analytes in the micelle aggregates; (ii) cloudiness; and (iii) phase separation for the analysis [48]. Nonionic surfactants can be able to form a micelle in aqueous solutions and become turbid at a specific temperature which is described as cloud point temperature. Over this point, micelle solution is divided into two phase: little volume phase that enriches in terms of surfactant and diluted aqueous phase. When metal ions react with an appropriate ligand, it can form aqueous low solubility complex, and therefore, these ions should be extracted from aqueous solution in the little volume enriched phase in terms of surfactant. This method is simple experimental procedure due to the low cost, eco-friendly, high capacity to preconcentration of the several analytes and good recovery with high enrichment factor. CPE is also simple, sensitive and rapid methods for concentration and separation of the essential elements [49].

CPE uses water and prevents the use of expensive, toxic and flammable organic solvents in a large volume. In addition to this, CPE should introduce several significant advantages such as faster operating, easy manipulation, short time, lower cost, higher recovery and enrichment factor, and less stringent requirements for the separation [50]. Diluted solvents of the surfactant can be used as an extractor media in the CPE, resulting in lower laboratory waste and cost-effective likely being economical reagents. Also, surfactants are less flammable than organic solvents [48].

4.9. Novel approaches in the field of solid phase extraction

Over the years, many new extraction techniques have been improved in food analysis. Selected applications involving extraction methods in food analysis are presented in **Table 1**. Recently, SPE has been improved according to the development of a simple and original device, which also serves as a magnetic stirrer [22]. Adsorptive μ-extraction (AμE) known as innovative extraction technique and its two versions (bar adsorptive μ-extraction (BAμE) and multispheres adsorptive μ-extraction (MSAμE)) were detailed. This technique used for determination of phenolic acid and triazines in some foods and beverages [81, 82]. Stir-rod-sorptive extraction (SRSE) device consists of a metallic wire with a magnet at one end which is the sorbent-coated glass [83]. SRSE allows extracting fluoroquinolones in honey [84]. Microsolid phase extraction (μ-SPE) was first developed in 2006. This new technique consists of positive features of SPE and capacity of membrane methods. A small bag (1–4 cm^2) contains adsorbent in its inside, and this bag is made of a porous membrane, and then, this bag should move freely in the sample or should be mixed in the sample headspace [22]. The main advantages of μ-SPE procedure are good analytical performance, reduced matrix effects, analysis time and solvent usage [85]. When compared to classical solid phase extraction, μ-SPE is more basic, more economical, more sensitive and less time-consuming process. Analytes are dissolved in a little solvent especially hexane and methanol after the extraction and then can be determined by GC or HPLC [22]. The μ-spe technique should be applied for the detection of biogenic amines in orange juice [77], and of organophosphorus pesticides from wheat [85]. Stir cake sorptive extraction (SCSE) is also another solvent free extraction methods and first reported in 2011. Monolithic cake of sorbent knowing as microporous material can be used as an extraction medium in SCSE, resulting in the high specific surface area. In the SCSE method, a special

Sample	Analyte	Selected sample preparation techniques	Ref
Madeira Wine	Aroma compounds	SPME/SBSE	[51]
Milk, egg, chicken and fish	Quinolone antibacterials	In-tube SPME	[52]
Water	Chlorophenols	HS-SPME	[53]
Tea infusions	Polycyclic aromatic hydrocarbons	HS-SPME	[54]
Wines, fruits, and vegetables	Famoxadone	DI-SPME	[55]
Food	Polycyclic aromatic hydrocarbons	SBSE	[56]
Juice	Pesticide residues and benzo[a]pyrene	SBSE	[57]
Vegetable	Organochlorine pesticides	SBSE	[58]
Bovine milk	Phenoxy herbicides	LLLME-hollow fiber membrane	[59]
Green tea leaves	Organochlorine pesticides	HF-LPME	[60]
Several foods	Nitrosamine	HF-LPME	[61]
Food	Phthalate esters	SDME	[39]
Orange juice	Organophosporus pesticide	SDME	[62]
Fruit juice and fruit extract	Organic acids	SDME	[63]
Water and wine	2, 4, 6-Tricholoroanisole	HS-IL-SDME	[64]
Egg yolks	Sudan dyes	MSPD–DLLME-	[65]
Grapes	Amino acids	UAE	[66]
Papaya seed oil	Physicochemical properties of papaya seed oil	UAE	[67]
Carrot	Carotenoids	UAE	[68]
Vegetables	Dichlorvos	MAE-HS-SPME	[69]
Meat products	Volatile nitrosamines	MAE-DLLME	[70]
Smoked fish	Polycyclic aromatic hydrocarbons	MAE-DLLME	[71]
Brazilian cherry seed	Phenolic compound	PFE	[72]
Scenedesmus obliquus	Carotenoids and chlorophylls a, b and c	SFE	[73]
Wine	Volatile components	SFE	[74]
Apple, green bean, carrot	Pesticide residue	SFE	[75]
Mineral water and drinkable water	Lead and cadmium	CPE	[76]
Orange juice	Biogenic amines	μ-SPE	[77]

Sample	Analyte	Selected sample preparation techniques	Ref
Water and food	Cobalt	Chelating agent free (CAF)-SPE	[78]
Food	Carotenoids	ASE	[79]
Bovine milk and dairy products	Nonsteroidal anti-inflammatory drugs	DLLME	[80]

Table 1. Selected applications of extraction techniques in food analysis.

holder consists of stationary phase and it has been developed to act as a magnetic stirrer; therefore, the holder has a significant role in both extractions of analytes and sample stirring [22]. SCSE method has some advantages such as basic operation, higher extraction capacity, cost-effectiveness, flexibility, and environmentally friendly. Monolith-based SCSE technique is applied to the juices such as soft drinks and milk, and honey to determine the sorbic acids, benzoic acid and cinnamic acid by Lin et al. [86] and to determine anthelmintics by Yulei et al. [87]. Polymeric ionic liquid-based SCSE is also used to determine organic preservatives in orange juice and tea drinks [88].

5. Conclusion

The preparation of food samples and preconcentration of analytes for the purpose of the analysis is necessary. Sample preparation is the main step in food analysis, greatly influencing the reliability and accuracy of results of analysis. Green chemistry approaches in the sample preparation techniques, as a sustainable and eco-friendly alternative to the classical techniques, are mandatory. At the same time, green sample preparation techniques are rapid, simple, generally solvent-free, sensitive, reliable and cost-effective. Different green microextraction techniques and its novel modifications have found an important role in sample preparation because of their inherent advantages over the conventional procedures. Modern trends in sample preparation techniques are toward the simplification and miniaturization of sample preparation, and the minimization of sample size and organic solvent used. In the forthcoming years, it is very probable that the greener techniques for sample preparation will be increasingly applied in food analysis, which is highly desirable.

Author details

Burak Demirhan[1], Hayriye Eda Şatana Kara[2] and Buket Er Demirhan[1]*

*Address all correspondence to: erbuket@gmail.com

1 Department of Food Analysis, Faculty of Pharmacy, Gazi University, Ankara, Turkey

2 Department of Analytical Chemistry, Faculty of Pharmacy, Gazi University, Ankara, Turkey

References

[1] Farajzadeh MA, Khoshmaram L, Sheykhizadeh S. A review on application of microextraction techniques for analysis of chemical. Analytical and Bioanalytical Chemistry Research. 2014;**1**(1):1-19

[2] Mendiola JA, Herrero M, Cifuentes A, Ibanez E. Use of compressed fluids for sample preparation: Food applications. Journal of Chromatography A. 2007;**1152**:234-246. DOI: 10.1016/j.chroma.2007.02.046

[3] Anastas PT, Kirchhoff MM, Williamson TC. Catalysis as a foundational pillar of green chemistry. Applied Catalysis A: General. 2001;**221**:3-13. PII: S0926-860X(01)00793-1

[4] Gałuszka A, Migaszewski Z, Namieśnik J. The 12 principles of green analytical chemistry and the significance mnemonic of green analytical practices. Trends in Analytical Chemistry. 2013;**50**:78-84. DOI:10.1016/j.trac.2013.04.010

[5] Armenta S, Garrigues S, Guardia M. The role of green extraction techniques in Green Analytical Chemistry. Trends in Analytical Chemistry. 2015;**71**:2-8. DOI: 10.1016/j.trac.2014.12.011

[6] Tiwari BK. Ultrasound: A clean, green extraction technology. Trends in Analytical Chemistry. 2015;**71**:100-109. DOI: 10.1016/j.trac.2015.04.013

[7] Anastas P, Eghbali N. Green chemistry: Principles and practice. Chemical Society Reviews. 2010;**39**:301-312. DOI: 10.1039/b918763b

[8] Warner JC. Cannon AS, Dye KM. Green chemistry. Environmental impact assessment: Review. 2004;**24**:775-799. DOI: 10.1016/j.eiar.2004.06.006

[9] Armenta S, Garrigues S, Guardia M. Green analytical chemistry. Trends in Analytical Chemistry. 2008;**27**(6):497-511. DOI:10.1016/j.trac.2008.05.003

[10] Anastas PT, Bartlett LB, Kirchhoff MM, Williamson TC. The role of catalysis in the design, development, and implementation of green chemistry. Catalysis Today. 2000;**55**:11-22. PII: S0920-5861(99)00222-9

[11] Anastas PT, Warner JC. In Green Chemistry: Theory and Practice. New York: Oxford University Press; 1998

[12] Horvath I and Anastas PT. Introduction: Green Chemistry. Chemical Reviews. 2007;**107**:2167-2168. DOI: 10.1021/cr0783784

[13] Płotka J, Tobiszewski M, Sulej AM, Kupska M, Górecki T, Namiesnik J. Green chromatography. Journal of Chromatography A. 2013;**1307**:1-20. DOI: 10.1016/j.chroma.2013.07.099

[14] Kataoka H, Lord HL, Pawliszyn J. Review: Applications of solid-phase microextraction in food analysis. Journal of Chromatography A. 2000;**880**:35-62. PII: S0021-9673(00)00309-5

[15] Hailemariam T, Bekele T. Recent advances in sample preparation techniques for environmental matrix. International Journal of Modern Chemistry and Applied Science. 2015;**2**(2):92-107

[16] Fumes BH, Silva MR, Andrade FN, Nazario CED, Lanças FM. Recent advances and future trends in new materials for sample preparation. Trends in Analytical Chemistry. 2015;**71**:9-25. DOI: 10.1016/j.trac.2015.04.011

[17] Moliner-Martinez Y, Herráez-Hernández R, Verdú-Andrés J, Molins-Legua C,Campíns-Falcó P. Recent advances of in-tube solid-phase microextraction. Trends in Analytical Chemistry. 2015;**71**:205-213. DOI: 10.1016/j.trac.2015.02.020

[18] Er Demirhan B, Demirhan B, Baş Yarımkaya S, Yentür G, Öktem AB. Investigation of melamine presence in canned tuna fish by capillary zone electrophoresis method. Journal of Agricultural Sciences. 2015;**21**:310-315

[19] Kinsella B, O'Mahony J, Malone E, Moloney M, Cantwell H, Furey A, Danaher M. Current trends in sample preparation for growth promoter and veterinary drug residue analysis. Journal of Chromatography A. 2009;**1216**(46):7977-8015. DOI: 10.1016/j.chroma.2009.09.005

[20] Beyer A, Biziuk M. Applications of sample preparation techniques in the analysis of pesticides and PCBs in food. Food Chemistry. 2008;**108**:669-680. DOI:10.1016/j.foodchem.2007.11.024

[21] Andrašcíková M, Matisová E, Hrouzková S. Liquid phase microextraction techniques as a sample preparation step for analysis of pesticide residues in food. Separation & Purification Reviews. 2015;**44**:1-18. DOI: 10.1080/15422119.2013.872125

[22] Spietelun A, Marcinkowski L, Guardia M, Namiesnik J. Review: Recent developments and future trends in solid phase microextraction techniques towards green analytical chemistry. Journal of Chromatography A. 2013;**1321**:1-13 DOI: 10.1016/j.chroma.2013.10.030

[23] Arthur CL, Pawliszyn J. Solid phase microextraction with thermal desorption using fused silica optical fibers. Analytical Chemistry. 1990;**62**(19):2145-2148. DOI: 10.1021/ac00218a019

[24] Vas G, V′ekey K. Solid-phase microextraction: A powerful sample preparation tool prior to mass spectrometric analysis. Journal of Mass Spectrometry. 2004;**39**:233-254. DOI: 10.1002/jms.606

[25] Curyło J, Wardencki W, Namieśnik J. Green aspects of sample preparation—a need for solvent reduction. Polish Journal of Environmental Studies. 2007;**16** 1:5-16

[26] Kumar A, Gaurav, Malik AK, Tewary DK, Singh B. A review on development of solid phase microextraction fibers by sol-gel methods and their applications. Analytica Chimica Acta. 2008;**610**:1-14. DOI: 10.1016/j.aca.2008.01.028

[27] Wilkes JG, Conte ED, Kim Y, Holcomb M, Sutherland JB, Miller DW. Review: Sample preparation for the analysis of flavors and off-flavors in foods. Journal of Chromatography A. 2000;**880**:3-33. PII: S0021-9673(00)00318-6

[28] Kataoka H. Automated sample preparation using in tube solid phase microextraction and its application—A review. Analytical and Bioanalytical Chemistry. 2002;**373**(1):31-45. DOI: 10.1007/s00216-002-1269-z

[29] Pawliszyn J. Theory of solid-phase microextraction. Journal of Chromatographic Science. 2000;**38**:270-278

[30] Merkle S, Kleeberg KK, Fritsche J. Recent developments and applications of solid phase microextraction (SPME) in food and environmental analysis—A review. Chromatography. 2015;**2**:293-381. DOI: 10.3390/chromatography2030293

[31] Baltussen E, Sandra P, David F, Cramers C. Stir bar sorptive extraction (SBSE), a novel extraction technique for aqueous samples: Theory and principles. Journal of Microcolumn Separations. 1999;**11**:737-747. DOI: 10.1002/(SICI)1520667X(1999)11:10<737::AIDMCS7>3.0.CO;24

[32] Nogueira JMF. Stir-bar sorptive extraction: 15 years making sample preparation more environment-friendly. Trends in Analytical Chemistry. 2015;**71**:214-223. DOI: 10.1016/j.trac.2015.05.002

[33] Chen Y, Guo Z, Wang X, Qiu C. Review: Sample preparation. Journal of Chromatography A. 2008;**1184**:191-219. DOI: 10.1016/j.chroma.2007.10.026

[34] Yu C, Liu Q, Lan L, Hu B. Comparison of dual solvent-stir bars microextraction and U-shaped hollow fiber-liquid phase microextraction for the analysis of Sudan dyes in food samples by high-performance liquid chromatography-ultraviolet/mass spectrometry. Journal of Chromatography A. 2008;**1188**:124-131. DOI: 10.1016/j.chroma.2008.02.065

[35] Alsharif AMA, Huat TG, Mun CY, Lawal A. Liquid phase microextraction for analysis of mycotoxins in food samples: Review. Research Journal of Chemical and Environmental Sciences. 2015;**13**:05-21

[36] Sarafraz-Yazdi A, Amiri A. Liquid-phase microextraction. Trends in Analytical Chemistry. 2010;**29**(1):1-14. DOI: 10.1016/j.trac.2009.10.003

[37] Liu H, Dasgupta PK. Analytical chemistry in a drop. Solvent extraction in a microdrop. Analytical Chemistry. 1996;**1, 68**(11):1817-1821. DOI: 10.1021/ac960145h

[38] Kokosa JM. Recent trends in using single-drop microextraction and related techniques in green analytical methods. Trends in Analytical Chemistry. 2015;**71**:194-204. DOI: 10.1016/j.trac.2015.04.019

[39] Li M, Li Y, Mao L. Application of single drop microextraction in the determination of phthalate esters in food by gas chromatography mass spectrometry. Se Pu Chinese Journal of Chromatography. 2007;**25**(1):35-38

[40] Rezaeea M, Assadia Y, Hosseinia MRM, Aghaeea E, Ahmadia F, Berijani S. Determination of organic compounds in water using dispersive liquid–liquid microextraction. Journal of Chromatography A. 2006;**1116**(1-2):1-9. DOI: 10.1016/j.chroma.2006.03.007

[41] Bjergaard SP, Rasmussen KE. Liquid-Liquid-Liquid microextraction for sample preparation of biological fluids prior to capillary electrophoresis. Analytical Chemistry. 1999;**71**:2650-2656. DOI: 10.1021/ac990055n

[42] Rasmussen KE, Bjergaard SP. Developments in hollow fibrebased, liquid-phase microextraction. Trends in Analytical Chemistry. 2004;**23**(1):1-10. DOI: 10.1016/S0165-9936(04)00105-0

[43] Pena-Pereira F, Lavilla I, Bendicho C. Miniaturized pre-concentration methods based on liquid–liquid extraction and their application in inorganic ultra-trace analysis and speciation: A review. Spectrochimica Acta Part B. 2009;**64**:1-15. DOI:10.1016/j.sab.2008.10.042

[44] Ridgway K, Lalljie SPD, Smith RM. Review: Sample preparation techniques for the determination of trace residues and contaminants in foods. Journal of Chromatography A. 2007;**1153**:36-53. DOI: 10.1016/j.chroma.2007.01.134

[45] Tiwari BK. Ultrasound: A clean, green extraction technology. Trends in Analytical Chemistry. 2015;**71**:100-109. DOI: 10.1016/j.trac.2015.04.013

[46] Chemat F, Rombaut N, Sicaire AG, Meullemiestre A, Tixier ASF, Vian MA. Ultrasound assisted extraction of food and natural products. Mechanisms, techniques, combinations, protocols and applications: A review. Ultrasonics Sonochemistry. 2017;540-560. DOI: 10.1016/j.ultsonch.2016.06.035

[47] Koning S, Janssen HG, Brinkman UAT. Modern methods of sample preparation for GC analysis. Chromatographia Supplement. 2009;**69**:s33-s78. DOI: 10.1365/s10337-008-0937-3

[48] Sahin CA, Efecınar M, Satıroglu N. Combination of cloud point extraction and flame atomic absorption spectrometry for pre-concentration and determination of nickel and manganese ions in water and food samples. Journal of Hazardous Materials. 2010;**176**:672-677. DOI: 10.1016/j.jhazmat.2009.11.084

[49] Wang S, Meng S, Guo Y. Cloud point extraction for the determination of trace amounts of cobalt in water and food samples by flame atomic absorption spectrometry. Journal of Spectroscopy. 2013;**735702**:1-7. DOI: 10.1155/2013/735702

[50] Citak D, Tuzen M. A novel pre-concentration procedure using cloud point extraction for determination of lead, cobalt and copper in water and food samples using flame atomic absorption spectrometry. Food and Chemical Toxicology. 2010;**48**:1399-1404. DOI: 10.1016/j.fct.2010.03.008

[51] Alves RF, Nascimento AMD, Nogueira JMF. Characterization of the aroma profile of Madeira wine by sorptive extraction techniques. Analytica Chimica Acta. 2005;**546**:11-21. DOI: 10.1016/j.aca.2005.05.012

[52] Zheng MM, Ruan GD, Feng YQ. Evaluating polymer monolith in-tube solid-phase microextraction coupled to liquid chromatography/quadrupole time-of-flight mass spectrometry for reliable quantification and confirmation of quinolone antibacterials in edible animal food. Journal of Chromatography A. 2009;**1216**:7510-7519. DOI: 10.1016/j.chroma.2009.03.054

[53] Holopainen S, Luukkonen V, Nousiainen M, Sillanpää M. Determination of chlorophenols in water by headspace solid phase microextraction ion mobility spectrometry (HS-SPME-IMS). Talanta. 2013;**114**:176-182. DOI: 10.1016/j.talanta.2013. 04.023

[54] Vinas P, Campillo N, Aguinaga N, Pérez-Cánovas E, Hernández-Córdoba M. Use of headspace solid-phase microextraction coupled to liquid chromatography for the analysis of polycyclic aromatic hydrocarbons in tea infusions. Journal of Chromatography A. 2007;**1164**:10-17. DOI: 10.1016/j.chroma.2007.06.056

[55] Aguinaga N, Campillo N, Viñas P, Hernández-Córdoba M. Solid phase microextraction coupled to gas chromatography–mass spectrometry for the analysis of famoxadone in wines, fruits, and vegetables. Spectroscopy Letters. 2009;**42**:320-326. DOI: 10.1080/003870 10903178723

[56] Zhang W, Zhang Z, Zhang J, Meng J, Bao T, Chen Z. Covalent immobilization of graphene onto stainless steel wire for jacket-free stir bar sorptive extraction. Journal of Chromatography A. 2014;**1351**:12-20. DOI: 10.1016/j.chroma.2014.05.038

[57] Zuin VG, Schellin M, Montero L, Yariwake JH, Augusto F, Popp P. Comparison of stir bar sorptive extraction and membrane-assisted solvent extraction as enrichment techniques for the determination of pesticide and benzo[a]pyrene residues in Brazilian sugarcane juice. Journal of Chromatography A. 2006;**1114**:180-187. DOI: 10.1016/j.chroma.2006.03.035

[58] Barriada-Pereira M, Serôdio P, González-Castro MJ, Nogueira JMF. Determination of organochlorine pesticides in vegetable matrices by stir bar sorptive extraction with liquid desorption and large volume injection-gas chromatography-mass spectrometry towards compliance with European Union directives. Journal of Chromatography A. 2010;**1217**:119-126. DOI: 10.1016/j.chroma.2009.10.076

[59] Zhu L, Ee KH, ZhaoL, Lee HK. Analysis of phenoxy herbicides in bovine milk by means of liquid-liquid-liquid microextraction with a hollow-fiber membrane. Journal of Chromatography A. 2002;**963**:335-343. PII: S0021-9673(02)00138-3

[60] Huang SP, Huang SD. Dynamic hollow fiber protected liquid phase microextraction and quantification using gas chromatography combined with electron capture detection of organochlorine pesticides in green tea leaves and ready-to-drink tea. Journal of Chromatography A. 2006;**1135**:6-11. DOI: 10.1016/j.chroma.2006.09.027

[61] Raharjo Y, Handajani US, Khasanah M. The application of hollow fiber-liquid phase microextraction based on green chemistry to analyze carcinogenic nitrosamines in food samples. Journal of Chemical and Pharmaceutical Research. 2014;**6**(5):1274-1280

[62] Zhao E, Han L, Jiang S, Wang Q, Zhou Z. Application of a single-drop microextraction for the analysis of organophosphorus pesticides in juice. Journal of Chromatography A. 2006;**1114**:269-273. DOI: 10.1016/j.chroma.2006.03.011

[63] Saraji M, Mousavinia F. Singledrop microextraction followed by in syringe derivatization and gas chromatography mass spectrometric detection for determination of organic acids in fruits and fruit juices. Journal of Separation Science. 2006;**29**(9):1223-1229

[64] Márquez-Sillero I, Aguilera-Herrador E, Cárdenas S, Valcárcel M. Determination of 2, 4, 6-tricholoroanisole in water and wine samples by ionic liquid-based single-drop microextraction and ion mobility spectrometry. Analytica Chimica Acta. 2011;**702**:199-204. DOI: 10.1016/j.aca.2011.06.046

[65] Yan H, Wang H, Qiao J, Yang G. Molecularly imprinted matrix solid-phase dispersion combined with dispersive liquid–liquid microextraction for the determination of four Sudan dyes in egg yolk. Journal of Chromatography A. 2011;**1218**:2182-2188. DOI: 10.1016/j.chroma.2011.02.042

[66] Carrera C, Ruiz-Rodríguez A, Palma M, Barroso CG. Ultrasound-assisted extraction of amino acids from grapes. Ultrasonics Sonochemistry. 2015;**22**:499-505. DOI: 10.1016/j.ultsonch.2014.05.021

[67] Samaram S, Mirhosseini H, Tan CP, Ghazali HM. Ultrasound-assisted extraction and solvent extraction of papaya seed oil: Crystallization and thermal behavior, saturation degree, color and oxidative stability. Industrial Crops and Products. 2014;**52**:702-708. DOI: 10.1016/j.indcrop.2013.11.047

[68] Li Y, Fabiano-Tixier AS, Tomao V, Cravotto G, Chemat F. Short Communication: Green ultrasound-assisted extraction of carotenoids based on the bio-refinery concept using sunflower oil as an alternative solvent. Ultrasonics Sonochemistry. 2013;**20**:12-18. DOI: 10.1016/j.ultsonch.2012.07.005

[69] Chen YI, Su YS, Jen JF. Determination of dichlorvos by on-line microwave-assisted extraction coupled to headspace solid-phase microextraction and gas chromatography–electron-capture detection. Journal of Chromatography A. 2002;**976**:349-355. P II: S0021-9673(02)01144-5

[70] Campillo N, Vinas P, Martínez-Castillo N, Hernández-Córdoba M. Determination of volatile nitrosamines in meat products by microwave-assisted extraction and dispersive liquid–liquid microextraction coupled to gas chromatography–mass spectrometry. Journal of Chromatography A. 2011;**1218**:1815-1821. DOI: 10.1016/j.chroma.2011.02.010

[71] Ghasemzadeh-Mohammadi V, Mohammadi A, Hashemi M, Khaksar R, Haratian P. Microwave-assisted extraction and dispersive liquid–liquid microextraction followed by gas chromatography–mass spectrometry for isolation and determination of polycyclic aromatic hydrocarbons in smoked fish. Journal of Chromatography A. 2012;**1237**:30-36. DOI: 10.1016/j.chroma.2012.02.078

[72] Oliveira AL, Destandau E, Fougère L, Lafosse M. Isolation by pressurised fluid extraction (PFE) and identification using CPC and HPLC/ESI/MS of phenolic compounds from Brazilian cherry seeds (*Eugenia uniflora* L.). Food Chemistry. 2014;**145**:522-529. DOI: 10.1016/j.foodchem.2013.08.065

[73] Guedes AC, Gião MS, Matias AA, Nunes AVM, Pintado ME, Duarte CMM, Malcata FX. Research note: Supercritical fluid extraction of carotenoids and chlorophylls a, b and c, from a wild strain of *Scenedesmus obliquus* for use in food processing. Journal of Food Engineering. 2013;**116**:478-482. DOI: 10.1016/j.jfoodeng.2012.12.015

[74] Karasek P, Planeta J, Ostra EV, Mikesova M, Golias J, Roth M, Vejrosta J. Direct continuous supercritical fluid extraction as a novel method of wine analysis Comparison with conventional indirect extraction and implications for wine variety identification. Journal of Chromatography A. 2003;**1002**:13-23. DOI: 10.1016/S0021-9673(03)00561-2

[75] Lehotay SJ. Determination of pesticide residues in nonfatty foods by supercritical fluid extraction and gas chromatography/mass spectrometry: Collaborative study. Journal of AOAC International. 2002;**85**(5):1148-1166

[76] Silva EL, Roldan PS. Simultaneous flow injection pre-concentration of lead and cadmium using cloud point extraction and determination by atomic absorption spectrometry. Journal of Hazardous Materials. 2009;**161**:142-147. DOI: 10.1016/j.jhazmat.2008.03.100

[77] Basheer C, Wong W, Makahleh A, Tameem AA, Salhin A, Saad B, Lee HK. Hydrazone-based ligands for micro-solid phase extraction-high performance liquid chromatographic determination of biogenic amines in orange juice. Journal of Chromatography A. 2011;**1218**:4332-4339. DOI: 10.1016/j.chroma.2011.04.073

[78] Diniz KM, Gorla FA, Ribeiro ES, Nascimento MBO, Corrêa RJ, Tarley CRT, Segatelli MG. Preparation of $SiO_2/Nb_2O_5/ZnO$ mixed oxide by sol–gel method and its application for adsorption studies and on-line pre-concentration of cobalt ions from aqueous medium. Chemical Engineering Journal. 2014;**239**:233-241. DOI: 10.1016/j.cej.2013.11.027

[79] Breithaupt DE. Analytical, nutritional and clinical methods simultaneous HPLC determination of carotenoids used as food coloring additives: Applicability of accelerated solvent extraction. Food Chemistry. 2004;**86**:449-456.DOI: 10.1016/j.foodchem.2003.10.027

[80] Alshana U, Göğer NG, Ertaş N. Dispersive liquid–liquid microextraction combined with fieldamplified sample stacking in capillary electrophoresis for the determination of nonsteroidal anti-inflammatory drugs in milk and dairy products. Food Chemistry. 2013;**38**(2-3):890-897. DOI: 10.1016/j.foodchem.2012.11.121

[81] Neng NR, Sequeiros RCP. Combining bar adsorptive microextraction with capillary electrophoresis—Application for the determination of phenolic acids in food matrices. Electrophoresis. 2014;**35**(17):2488-2494. DOI: 10.1002/elps.201400101

[82] Andrade FN, Ide AH, Neng Nda R, Lanças FM, Nogueira JM. Determination of trace levels of triazines in corn matrices by bar adsorptive microextraction with a molecularly imprinted polymer. Journal of Separation Science. 2016;**39**(4):756-761. DOI: 10.1002/jssc.201501101

[83] Namieśnik J, Spietelun A, Marcinkowski L. Green sample preparation techniques for chromatographic determination of small organic compounds. International Journal of Chemical Engineering and Applications. 2015;**6**(3):215-219. DOI: 10.7763/IJCEA.2015.V6.484

[84] Luo YB, Ma Q, Feng YQ. Stir rod sorptive extraction with monolithic polymer as coating and its application to the analysis of fluoroquinolones in honey sample. Journal of Chromatography A. 2010;**1217**:3583-3589. DOI: 10.1016/j.chroma.2010.03.036

[85] Pelle FD, Di Crescenzo MC, Sergi M, Montesano C, Di Ottavio F, Scarpone R. Micro-solid-phase extraction (μ-SPE) of organophosphorous pesticides from wheat followed by LCMS/MS determination. Food Additives and Contaminants: Part A. 2016;**33**:291-299. DOI: 10.1080/19440049.2015.1123818

[86] Lin F, Nong S, Huang X, Yuan D. Sensitive determination of organic acid preservatives in juices and soft drinks treated by monolith-based stir cake sorptive extraction and liquid chromatography analysis. Analytical and Bioanalytical Chemistry. 2013;**405**:2077-2081. DOI: 10.1007/s00216-012-6646-7

[87] Yulei W, Jie Z, Xiaojia H, Dongxing Y. Preparation of stir cake sorptive extraction based on polymeric ionic liquid for the enrichment of benzimidazole anthelmintics in water, honey and milk samples. Analytica Chimica Acta. 2014;**46**:33-41. DOI: 10.1016/j.aca.2014.06.039

[88] Chen L, Huang X. Preparation of a polymeric ionic liquid-based adsorbent for stir cake sorptive extraction of preservatives in orange juices and tea drinks. Analytica Chimica Acta. 2016;**916**:33-41. DOI: 10.1016/j.aca.2016.02.030

Chapter 2

Reproducible Sample Preparation for Reliable Food Analysis

Tanja Butt

Additional information is available at the end of the chapter

http://dx.doi.org/10.5772/intechopen.68811

Abstract

A reliable and accurate analysis of food samples can only be guaranteed by reproducible sample preparation. This chapter describes the process of turning a laboratory sample into a representative part sample with homogeneous analytical fineness by choosing the most suitable mill. Important aspects of size reduction and homogenization are explained, a variety of application examples is given, and specific applications such as cryogenic grinding are discussed in detail.

Keywords: sampling and sample division, sample preparation, homogenization, pulverization, milling, cryogenic grinding, minimizing standard deviation, particle size

1. Introduction

Food occurs in a great variety of consistencies and is often inhomogeneous. Food testing laboratories require representative samples to produce meaningful and reproducible analysis results. Therefore, food samples have to be homogenized and pulverized to the required analytical fineness, ideally with as little time and effort as possible. Furthermore, reliable analytical results can only be obtained if the entire sample preparation process is carried out reproducibly [1]. This means that the prepared part sample, from which usually only a few grams or milligrams are required for analysis, needs to represent the laboratory sample as well as the original sample from which the laboratory sample was extracted. An inhomogeneous sample does not represent the original material because some components may be overrepresented or missing altogether. Consequently, a homogeneous sample is the basis for reliable and representative analytical results. A good example to understand the importance of sample homogeneity is fat analysis of pizza. Only a few milligrams of pizza are required for analysis. Random sampling might result in a piece of mushroom or salami or cheese which would falsify the total fat content in the subsequent analysis (**Figure 1**). However, if the pizza is first reduced to coarse particles of <5 mm and then pulverized to fine particles of <0.5 mm, a homogeneous, representative analysis sample is obtained.

Only a homogenized analytical sample fully represents the initial sample and ensures reliable and reproducible results—independent from which spot the part sample is taken. The standard deviation of any subsequent analysis can be minimized drastically by particle size reduction and homogenization of the analytical sample. In the pizza example, the fat content was measured (**Figure 2**). The fat content varies in the pizza samples with particle sizes around 5 mm, whereas it is much more consistent in the homogenized samples. The standard deviation (SD) is reduced from 0.21 to 0.03% (relative SD from 2.10 to 0.35%).

Figure 1. From left to right: a whole pizza; sample after grinding to coarse particles of <5 mm; fully homogenized sample with particle sizes of <0.5 mm.

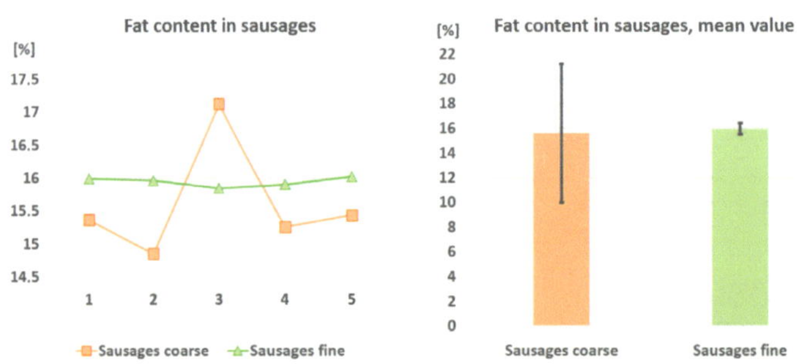

Figure 2. Left: fat content varies in coarse pizza samples but is stable in the pulverized samples; right: the mean values of each batch of five samples, the relative standard deviation of the fat content is reduced from 2.10 to 0.35% by homogenization.

2. How to select a suitable laboratory mill and accessories

When searching for a suitable mill and grinding tools, one should keep in mind that the sample properties to be determined (such as moisture or heavy metal content) must not be altered in any way during the process. To find the best suited mill for a specific application, the following aspects should be considered in advance:

- Feed size and required final fineness (Section 2.1).
- Characteristics of the sample and size reduction principles (Section 2.2).
- Sample volume and sample throughput (Section 2.3).
- Grinding tools and subsequent analysis (Section 2.4).
- Drying or embrittlement of the sample (Section 2.5).

2.1. Feed size and required final fineness

The feed size means the original particle size of the sample. For choosing a suitable mill, it makes a great difference whether large samples, like a whole fish, or small particles, such as crop grains, are to be homogenized. Whereas samples with small particle sizes can be fed directly into most grinders and mills, large-sized samples do not fit into every mill. Therefore, manual size reduction, for example, by cutting, or a preliminary grinding step in another mill may be required. The mills that accept larger initial particle sizes are mostly not suitable for producing very fine particles which are small enough for subsequent analysis. A frequent requirement is to "grind the sample to fine powder," but the term "powder" is not precise [2]. Washing powder, coffee powder, or baking powder have very different particle size distributions. Another typical request is to have the sample ground "as fine as possible." This involves a high input of energy and time which in turn increases costs. A much more effective approach is to grind the sample as fine as necessary. The required analytical fineness of the sample material depends on the analytical method or further processing and can vary greatly. Most methods require a fineness in the size range from 20 μm to 2 mm. As product properties (e.g., extraction, filtration, or absorption capacity) are often influenced by the particle size, size reduction on a laboratory scale is also essential for the development of new products or production processes.

2.2. Characteristics of the sample and size reduction principles

Depending on the sample properties, different size reduction principles are suitable to obtain the required fineness [3]. As mentioned before, large particles cannot always be ground to analytical fineness in one step. In some cases, it is possible to carry out preliminary and fine grinding in the same mill with different settings; in other cases, two mills are required. Another essential aspect relates to the sample properties: to produce a size reduction effect, the comminution principle of the mill should match the breaking behavior of the sample. Therefore, when selecting a suitable instrument, a thorough evaluation of the sample material is essential. Properties such as density, hardness, consistency, residual moisture, or fat content must be considered. Other characteristics, which may influence the success of the grinding process, are temperature stability or tendency of the sample to agglomerate.

Laboratory mills work with different size reduction principles. The type of mill used always depends on the breaking properties of the sample material. The subsequent pages show the most common mechanisms for the size reduction of solids. Usually, various size reduction principles are combined in one mill, such as impact and friction in planetary ball mills or shearing and impact in rotor mills. Hard and brittle samples are pulverized best by pressure, impact, and friction. These size reduction principles, however, have only limited effect on fibrous, soft, elastic samples. Pulverizing a freeze-dried fish, for example, is not possible by using pressure or impact; cutting and shearing are suited much better.

2.2.1. Size reduction of hard and brittle materials

Hard and brittle materials can be crushed with pressure, impact effects and/or friction (**Figure 3**). Pressure means a force which is applied between two solid surfaces that either represent the grinding tool surfaces directly or may be the surfaces of adjacent particles. Pressure is exerted by the grinding tools (jaw crushers, toggle crushers). Impact means a force at a solid surface. This could either be that of a grinding tool, or be represented by other particles. Strain by impact is mainly caused by one-sided and opposing particle acceleration (mixer mills, planetary mills, impact mills, jet impact mills, and drum mills). Friction means a force between two solid surfaces, caused by the vertical pressure of one surface and the simultaneous movement of the other surface (mortar grinders, disc mills, hand mortars, and rod mills).

2.2.2. Size reduction of soft, elastic, and fibrous materials

Shearing and cutting mechanisms are best suited to pulverize soft, elastic, and fibrous materials (**Figure 4**). Shearing means a force between two or more solid surfaces moving in opposite

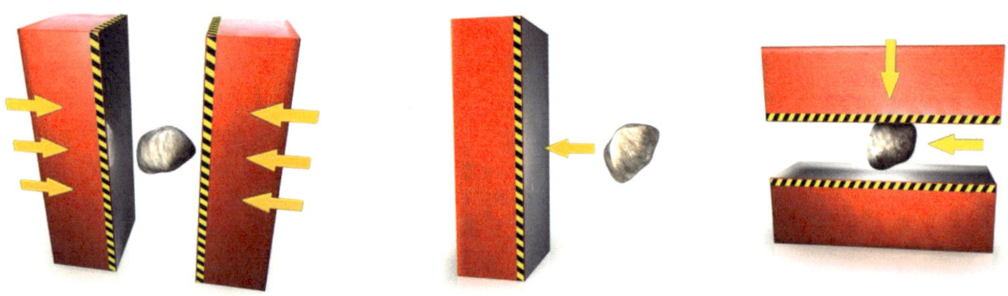

Figure 3. Size reduction principles for hard and brittle materials: pressure, impact, and friction.

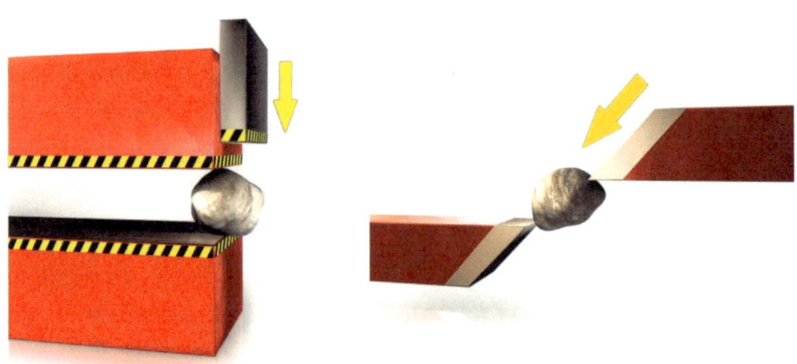

Figure 4. Size reduction principles for soft, elastic, fibrous materials: shearing and cutting.

directions (rotor beater mills, cross-beater mills, and ultra centrifugal mills). Cutting means a force between two or more sharpened surfaces (shredders, cutting mills, and knife mills).

2.3. Sample volume and sample throughput

Some mills accept larger sample quantities than others. Open systems with an inlet and outlet, like rotor mills, may be fed with larger amounts of bulk material. However, if the mill has a closed grinding chamber, for example, the grinding jar of a ball mill, the sample amount which can be processed in one batch is limited. Grinding kilogram quantities of wheat in a rotor mill is carried out much quicker than grinding the same amount in a mixer mill with a maximum jar volume of 50 mL (sample amount of 20 mL). The sample throughput may also influence the choice of a mill. It is based on the time required to assemble all parts of the mill, to grind the sample, and to clean the mill between two different samples. If only a few samples are analyzed each day, increased effort for a singular sample preparation process may be tolerable but not if hundreds of samples need to be prepared on a daily basis.

2.4. Grinding tools and subsequent analysis

For most mills and crushers, a variety of accessories and grinding tools are available. The selection of suitable accessories ensures effective grinding processes and reliable results. Two aspects should be considered: Which accessories are most effective and how the subsequent analysis might be influenced by the material of the grinding tools. Grinding tools are available in different materials, depending on the type of mill. The most common are the following:

- Metal (steel, cast iron, titanium).
- Ceramics (tungsten carbide, zirconium oxide, sintered aluminum oxide, hard porcelain, glass).
- Natural stone (agate).
- Plastics (PTFE, PC, PP, PE).

Grinding tools made of steel are available for all mills. When choosing a suitable grinding set, several factors must be considered, such as the hardness of the sample material and its breaking properties. The material of the grinding set should be harder than the sample to avoid excessive wear. In the case of food samples, this is true for most of the grinding materials used. Another important feature, which is mostly relevant for ball mills, is the energy input generated by the different materials. Grinding balls of tungsten carbide, for example, generate a much higher energy input, and thereby a better size reduction effect, due to the higher density of the material, than balls of the same size of other materials. On the other hand, too much crushing efficiency leads to caking of the sample material on the jar walls, especially in ball mills. This applies to soft, fatty, and sticky materials, characteristics which are typical for food samples. Therefore, the energy input must be considered carefully to avoid these effects.

Mechanical size reduction always leads to a certain degree of abrasion which may influence the subsequent analysis. Consequently, traces of materials like steel or zirconium oxide may be found in the sample. Anyhow, the amount is usually below detection limit for most analyses and can therefore be neglected. Moreover, some analyses, for example, determination of the fat content, are not affected by the iron and chromium traces resulting from steel abrasion. If, however, the heavy metal content is the object of investigation, the abrasion coming from steel equipment may lead to falsified results. In this case, using tools made of a "neutral" material like zirconium oxide or tungsten carbide is more advisable. The degree of abrasion also depends on the sample properties and the size reduction principle of the mill.

A special case is sample preparation under cryogenic conditions [4]: grinding with dry ice or liquid nitrogen should only be carried out with tools that are completely made of stainless steel. Plastic tools are not suitable as plastic embrittles at very low temperatures and may be damaged.

Regardless of the grinding tool material, the correct choice of accessories can have substantial influence on the grinding efficiency. For example, when grinding salad in a knife mill, the efficiency is greatly increased by using a gravity lid instead of a standard lid. Salad loses volume while being ground, and the gravity lid pushes the sample down against the knives for continuous homogenization.

2.5. Drying or embrittlement of the sample

2.5.1. Drying

It is only possible to grind moist or even wet sample materials without undesired side effects and sample loss with knife mills. When ground in rotor mills, moist materials tend to block the sieves which can lead to a blockage of the machine. As a consequence, material is lost and much time has to be spent on cleaning the mill. Therefore, it is advisable to dry the material before further processing. When choosing the drying method and temperature, care must be taken that the properties of the sample to be determined are not altered in any way. That is especially important with regard to temperature-sensitive or volatile components. Usually, these types of sample can only be air-dried at room temperature. Fluid bed dryers are suitable for gentle and quick drying of many products with an average drying time of 5–20 min. Further methods include vacuum and freeze drying as well as drying in ovens.

2.5.2. Embrittlement with liquid nitrogen or dry ice

Cooling the sample material often improves its breaking behavior. Some soft, tough, sticky, and fatty food materials have to be cooled before they can be subjected to preliminary or fine-size reduction. Chocolate or raisins, for example, can be pulverized easily by cryogenic grinding, whereas at room temperature, it is only possible to produce a paste with a low homogeneity. One way is to embrittle the sample in liquid nitrogen (LN_2) before grinding. At a temperature of −196°C, even soft jelly bears become so hard and brittle that they are pulverized without problems. Another possibility is to mix the sample with dry ice (solid CO_2). If the sample contains volatile substances which must be preserved for analysis, cryogenic grinding is also the method of choice. However, materials which must not become moist should not directly be treated with cooling agents, because the humidity of the air is condensing on the cold sample. Cooling agents should never be used in closed grinding tools as evaporation causes overpressure in the jar. More details of cryogenic grinding will be discussed at the end of this chapter.

3. Overview of mills commonly used for food sample preparation

Before specific application examples are discussed, we give an overview of the different mill types which are used most commonly for food sample preparation:

- Rotor mills (ultra centrifugal mills, cyclone mills, rotor beater mills).
- Knife mills.
- Ball mills (mixer mills, cryo mills).
- Cutting mills.

Obviously, more than one mill type may be suitable for grinding a particular sample, for example, wheat. As mentioned before, the choice of the most suitable mill for a certain sample depends on the sample volume, the required final fineness, the throughput, the material properties, and the subsequent analysis. The knowledge of the basics and working principles of different mill types helps to make the optimum choice for a specific application.

3.1. Rotor mills

Typical applications include seeds, corn, maize, wheat, dried algae, salt, sugar, dried fish, peas, nuts, almonds, coconut, coffee, tea, roots, gelatin, dried leaves, rice, spices, herbs, soya meal, and so on.

All types of rotor mills share the same grinding principle. The sample enters the mill through a hopper, hits on a rotor, which is either placed horizontally or vertically in the mill, and is smashed with impact onto the rotor teeth. In the second step, the sample passes a sieve with a specific aperture size. Here, mostly shearing effects are applied, with exception of the cyclone mill where friction prevails. Finally, the sample is collected in a bottle, cassette, or receptacle. In the following, three different types of rotor mills are discussed: ultra centrifugal mill, cyclone mill, and rotor beater mill.

3.1.1. Ultra centrifugal mills

Ultra centrifugal mills are used for the rapid fine-size reduction of soft, medium-hard, brittle, and fibrous materials. Size reduction is effected through impact and shearing forces between ring sieve and horizontal rotor. The maximum feed size is 10 mm. Especially with maximum speed, but depending on the material, a final fineness of 40 μm (d_{90}) and below may be achieved. Among the rotor mills, this is the highest achievable fineness. The grind size is determined by the aperture size of the exchangeable ring sieves (usually ranging from 0.08 to 10 mm). The revolution speed of ultra centrifugal mills ranges from 6000 to

18,000 min^{-1} or even more. The cassette principle guarantees 100% sample recovery and easy cleaning. It is recommendable to use a vibratory feeder for automatic and uniform feeding of large amounts of free-flowing materials. If large quantities or temperature-sensitive materials are processed, the use of a cyclone, for example, with a 3- or 5-L collector, is recommended. The frictional heat that is generated during the grinding process is partly discharged through the cyclone, so it helps to cool the sample. The use of distance sieves instead of standard ring sieves also helps to reduce frictional heat due to the greater gap between sieve plate and rotor. Accessories for ultra centrifugal mills usually include ring sieves and rotors of titanium for heavy-metal-free size reduction. If hard and abrasive materials are to be ground, a rotor with abrasion-resistant coating is required. For processing small amounts of sample, a mini-cassette with matching 316-L stainless steel rotor and various ring sieves is suitable.

Tips and techniques:

- When grinding an unknown sample, it is advisable to start with a sieve with a medium aperture size. The aperture size may be reduced if the sample does not block the sieve. This applies to all rotor mills.
- Grinding in ultra centrifugal mills is very effective: as a rule of thumb, 80% of the pulverized sample is smaller than half the aperture size of the sieve.
- If the sound of the machines changes significantly and/or if dusty material suddenly comes out of the hopper, the grinding chamber must be checked for overload or blocked sieves.
- When grinding temperature-sensitive materials, a cyclone helps to reduce the temperature. Distance sieves have the same effect.
- If the sample is fatty, the use of a distance sieve is advisable, as the shearing effect is reduced, and consequently less fat is "squeezed" from the particles which might block the apertures.
- Large particles should be pre-crushed using a sieve with medium to large aperture size. Fine grinding in a second step using a finer sieve is mostly quicker than trying to force large particles directly through the small apertures.
- If sample material remains in the grinding chamber although a cyclone is used, removing the sieve and letting the mill run for a few seconds clears the chamber. Repeat this step from time to time during milling of large quantities.

3.1.2. Cyclone mills

Cyclone mills are specially designed for the processing of foods and feedstuff for subsequent near-infrared spectroscopy (NIR analysis). They process fibrous and soft products quickly and gently to the required analytical fineness of about 0.5 mm. The mills are ideally suited for grinding various types of non-fatty food. They are equipped with a rotor and grinding ring with sieve insert. The high-revolution speed of up to 14,000 min^{-1} and the grinding geometry of the rotor and grinding chamber generate an air stream which carries the sample through the integrated cyclone into the sample bottle. This helps to avoid cross-contaminations. The cyclone provides additional cooling of the sample and the grinding tools. This prevents loss of moisture and thermal degradation ensuring preservation of the sample properties to be determined. The ground material is separated in the cyclone and collected in a sample bottle for complete recovery. The rotor speed can be adjusted in three steps allowing for perfect adaptation to sample requirements.

Tips and techniques:

- Quick exchange of sample bottles for increased throughput of samples.
- No cross-contamination and low cleaning effort required.
- As the grinding principle of cyclone mills has impact and friction, the machine should not be used for fatty sample materials like oil seeds.

3.1.3. Rotor beater mills

Rotor beater mills are used for the preliminary and fine-size reduction of soft, medium-hard, and brittle materials with a maximum feed size of 25 mm. The final fineness is determined by the aperture size of the exchangeable ring sieves (0.08–10 mm). A fineness down to 50 μm and below, depending on the properties of the sample material, may be achieved. Size reduction in the rotor beater mill is effected by impact and shearing forces between the vertical rotor and the ring sieve. To achieve an additional size reduction effect through impact, a 180°-grinding insert may be used for harder materials. The revolution speed is adjustable between 3000 and 10,000 min^{-1}. For larger sample quantities, a vibratory feeder can be used for automated feeding. In contrast to ultra centrifugal mills and cyclone mills, rotor beater mills are also suitable for grinding large sample amounts up to 30 L in one step.

Tips and techniques:

- A higher speed generates a higher throughput and less frictional heat.
- For temperature-sensitive materials, the use of a distance rotor is recommended. The larger grinding gap ensures a reduction of frictional heat.
- A cyclone is available which also reduces the heat build-up by discharging ground particles quickly out of the grinding chamber and generating a cooling air flow.

3.2. Knife mills

Typical applications: fresh meat, herbs, milk powder, fresh bacon, convenience food, cereal bars, soy beans, cakes, fresh fish, salad, cabbage, raisins, tomatoes, apples, fresh vegetables, sweets, jelly bears, bread, cheese, liver, fruits, chocolates, salami, soups, potatoes, cookies, waffles, ground beef, berries, nuts, seeds, boiled eggs, and so on.

Knife mills are suitable for the size reduction and homogenization of samples with a high fat, oil, or water content. They are frequently used in food control laboratories. The larger knife mills homogenize sample amounts up to 4500 mL, and are therefore the only mills which can homogenize a whole pizza or a loaf of bread in one batch. The speed range of the knife mills is flexible and allows for optimum adaption to the specific sample properties. When the mills are operated in reverse mode, the blunt end of the blades hits the sample with impact and crushes it (instead of cutting in a forward mode). A wide range of accessories are available: different knives and lids, containers of polypropylene, polycarbonate, stainless steel, and glass. Except for the polypropylene-grinding container, all containers can be autoclaved.

Tips and techniques:

- By using a gravity lid, the volume of the container is reduced and automatically adapted to the sample amount.
- For samples with a high liquid content, gravity lids with overflow channels are best suited. The liquid of the sample, which ascends the container walls, is returned to the center of the container for further homogenization.
- For heavy-metal-free grinding processes, neutral-to-analysis knives are available.
- By grinding in two or more steps (e.g., by changing from reverse to forward mode or by increasing the speed step by step), better grinding results may be achieved than by grinding in just one step.
- Always use the lid with two sealings when grinding wet samples in the larger-sized knife mills. Very wet samples must not be ground with maximum speed from beginning on, as the sample may splash out of the grinding container despite the sealings.

3.3. Ball mills

Typical applications include chocolate cream, spices, herbs, tea, olive pulp, lactose powder, egg shells, jelly bears, chitosan powder, liver, vanilla pods, berries, cookies, tobacco, chewing gum, wheat, waffles, frozen fish, seeds, and so on.

Ball mills are frequently used for the pulverization of hard-brittle materials. A crucial advantage of ball mills is their great versatility. Grinding jars and balls are available in various sizes and materials, for example, agate or ceramics such as zirconium oxide. This is important if the sample is analyzed for heavy metals. The grinding tools for ball mills consist of a grinding jar and grinding balls made of the same material. Mixer mills and cryo mills are the most widely used ball mills for homogenizing food samples.

Tips and techniques of ball mills in general:

- The following rule of thumb applies for the jar filling (dry grinding): one-third is filled with balls, one-third filled with sample material; thus, enough free space is left for ball movement. Also, this filling level ensures better grinding efficiency and less wear.

- When choosing the ball size, the feed size of the sample must be considered. For example, 30-mm grinding balls are suitable to grind particles of up to 10 mm.

3.4. Mixer mills

Mixer mills are suitable for grinding small sample quantities of up to 20 mL. The grinding jars perform radial oscillations in a horizontal position with a maximum frequency of 30 Hz. Size reduction is effected through impact forces, allowing for a final fineness down to d_{90} = 5 µm, depending on the sample properties. Grinding jars for mixer mills usually have a size range from 1.5 to 50 mL. When they are equipped with a screw-top lid, they are suitable for wet grinding. Another option is the use of different adapters which hold up to 20 × 2 or 10 × 5 mL reaction vials or 8 × 50 mL conical centrifuge tubes. Cryo mills are mixer mills specially designed for cryogenic grinding. They will be discussed later in the subchapter of cryogenic grinding.

Tips and techniques:

- If only one jar is filled with sample, the empty one should still be clamped to the second grinding station for balancing reasons.
- Closed grinding jars of steel can be cooled in liquid nitrogen to embrittle the sample material. Take care to fill in sample and grinding balls before cooling! Liquid nitrogen or dry ice must never be filled into the grinding jar—this would lead to overpressure inside the jar.
- Don't use grinding jars with mixed materials (e.g., steel jar with zirconium inlet) as the materials may react differently to very cold temperatures, leading to stress inside the jars and possibly to damages of the inlet.

3.5. Cutting mills

Typical applications include roots, tea, corn, freeze-dried fish, bones, mushrooms, spices, orange peel, sugar beet pellets, shea nuts, sugar cane, herbs, potatoes, lumps of cocoa butter, and so on.

Cutting mills are used for preliminary size reduction of soft, medium-hard, or fibrous materials such as roots, nut shells, or bones. Depending on the model, the revolution speed of the cutting mill is fixed or variable up to 3000 min^{-1}. The achievable grind size depends on the aperture size of the exchangeable bottom sieve (ranging from 0.25 to 20 mm) and the breaking properties of the sample material. Three types of rotors are available to find the best way to crush a specific sample: a parallel section rotor, acting like an axe, which is especially suitable for soft, elastic, and fibrous materials; a six-disc rotor with replaceable and reversible tungsten carbide plates, acting like a shredder, which is especially suitable for medium-hard materials; and finally, the V-rotor, acting like scissors, which is especially suitable for tough, soft, and fibrous material, improving the grinding process and sample discharge.

Tips and techniques:

- A cyclone helps to discharge the sample from the grinding chamber much quicker and leads to a cooling effect thanks to the generated air stream.
- Choose the most suitable rotor for a sample. The rotors either act like a shredder, an axe, or like scissors.

4. Application examples: homogenization of food

4.1. Fat content in sausages (knife mill)

Sausages often contain large fatty particles. They need to be thoroughly homogenized to ensure reliable analysis results. If the few grams required for fat content analysis were picked randomly from the sample, this would result in increased standard deviations of the analysis results. Two hundred grams of sausages was ground in two steps. After the sausages were cut manually into pieces of approximately 20 mm, a first grinding cycle was carried out in a knife mill at a revolution speed of 10,000 min^{-1} using a knife with serrated blades. The sample was cut to pieces smaller than 5 mm in only 15 s. The serrated blades help to tear the fibrous meat. A part sample was taken directly for fat analysis. The remaining sample was pulverized under cryogenic conditions. For this purpose, the sample was mixed with dry ice snow (with a ratio of 1:2) after the first grinding step and the mixture was then filled into a grinding container of stainless steel. Using a full metal knife and a lid specifically designed for cryogenic grinding, the sample was pulverized by grinding at 4,000 min^{-1} for 3 × 10 s (**Figure 5**).

Both the coarse and the homogenized samples were analyzed for their fat content five times by microwave-induced drying combined with nuclear magnetic resonance (NMR) spectroscopy. For each measurement, 4 g of sample was dried for 2.5 min and analyzed within 1 min. The fat content of the independent samples of the coarse sausage varies more than that of the finely ground samples. The fat content of the coarser fraction was measured in a range from 14.85 to 17.12% with a standard deviation (SD) of 0.88%. The SD was reduced by more than a factor of 10–0.07% in the homogenized sample (**Figure 6**), with a fat content ranging from 15.84 to 16.02% (relative SD reduced from 5.63 to 0.45%).

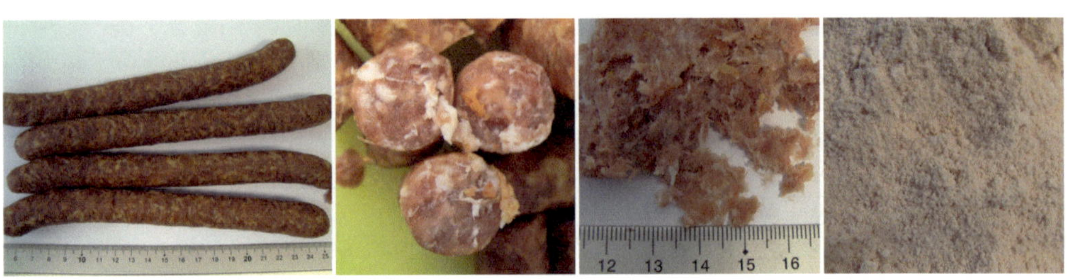

Figure 5. Homogenization of sausages; from left to right: original sample; pre-cut sample with large fatty parts; sample ground to <5 mm; pulverized sample of <300 μm.

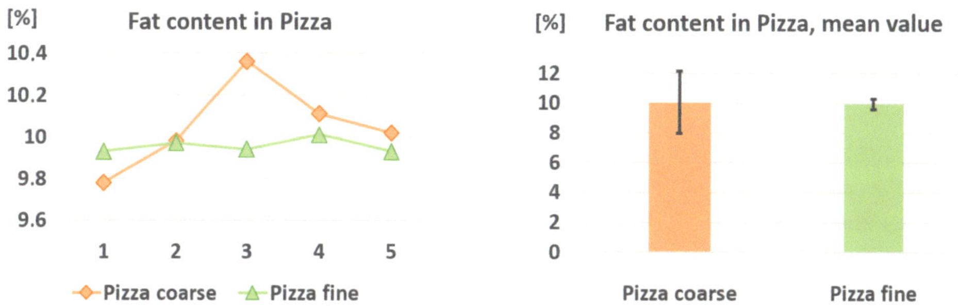

Figure 6. Left: fat content varies in coarse sausage samples, but is stable in fine-ground samples; right: the mean values of five samples of each of the relative standard deviations of the fat content in sausages were reduced from 5.63 to 0.45% by fine grinding.

4.2. NIR analysis of wheat samples (cyclone mill)

NIR is a common analytical method for the determination of protein content, moisture, fat, and ash in one run. Therefore, it is used whenever a high-sample throughput and great flexibility are required. A much-discussed issue is the necessity of sample preparation. What are the advantages of sample preparation before NIR analysis? The penetration depth of NIR radiation is 1 mm maximum, so everything that lies beneath cannot be detected. That is not a problem if the sample is completely homogeneous, but if a sample consists of different layers, like grains or seeds, then only the layers down to 1 mm are analyzed and are consequently overrepresented in the measurement results. To demonstrate this effect, the different properties of ground and unground wheat samples were analyzed with NIR [5]. The samples were analyzed 10 times, and the spectrometer was refilled for every measurement. The samples were pulverized in a cyclone mill. Cyclone mills are suitable for processing a variety of different materials which is ideal for NIR analysis requirements. The results for wheat show a large discrepancy between ground and unground sample, especially regarding the ash and fiber content (**Table 1**). As explained above, only the surface of the unground wheat is analyzed resulting in an overrepresentation of the kernel shell. Meaningful and reliable analysis results are guaranteed only by sample homogenization.

	Ash	Moisture	Fiber	Fat	Protein
Ground wheat					
Average [%]	2.80	9.68	1.10	1.17	9.02
Standard deviation [%]	0.03	0.09	0.05	0.03	0.07
Unground wheat					
Average [%]	0.10	9.80	6.90	1.38	8.46
Standard deviation [%]	0.10	0.25	0.62	0.16	0.45

Table 1. The analysis of wheat shows a difference in the ash and fiber content of the ground and unground sample.

4.3. Detection of mycotoxins in nuts (cutting mill and ultra centrifugal mill)

Mycotoxins are natural metabolism products of molds which have a toxic effect on humans and animals. Some types of food show an increased risk of mycotoxin release due to fungal infestation, especially when food is stored too long and in an unsuitable way. Fungal infestation usually occurs in nests, a random sample taken from the bulk must be sufficiently large to allow for the detection of contaminants. The first step is the preliminary size reduction of a representative amount of 1–2 kg per ton of nuts with a cutting mill to particles of <3 mm by using a bottom sieve of 4 mm [6]. It is important to use a six-disc rotor, as the shells of the nuts are too hard for the cutting effect of the other rotors. The subsequent fine-size reduction is ideally carried out with an ultra centrifugal mill. For the processing of hazelnuts, the use of distance sieves is recommended. As mycotoxins are lipophilic, the grinding process should be as gentle as possible to avoid the release of fat from the sample. A fineness of 300 μm (**Figure 7**) is sufficient for the subsequent extraction of the mycotoxins and for high-performance liquid chromatography (HPLC) analysis.

Figure 7. Homogenization of nuts; from left to right: original sample; pulverized sample of <300 μm.

4.4. Detection of amino acids in fatty, fresh bacon (knife mill)

Tough sample materials like fatty, inhomogeneous, streaky bacon pose a challenge to the homogenization process prior to analysis [7]. If larger parts of the rind or skin remain uncut, the sample is not homogeneous and the analysis may yield false results. Knife mills have proven to be best suited for thoroughly homogenizing meat samples (**Figure 8**). A strong motor to make use of the full cutting capacity of the blades is beneficial. Serrated blade knifes are ideally suitable for homogenizing tough meat samples in a very short time, as an additional tearing effect facilitates size reduction of the meat fibers. Short grinding times ensure low heat build-up. To obtain a thoroughly homogenized sample (at room-temperature conditions), the grinding process may require two or three steps. Two hundred and fifty grams of pork shoulder is processed in a knife mill with interval mode at a revolution speed of 3000 min^{-1}, using a serrated blade knife for 30 s. The first step is followed by two cycles of 30 s, each at 7000 min^{-1}. The best homogenization of the sample is achieved after another 30 s at 10,000 min^{-1}. The sample would bounce too much if the maximum speed was selected right from the start. Nonetheless, full speed is required at some point to achieve the best possible results. It is also important to use a standard lid, as other lid types might put too much pressure on the sample. The sample parts sticking to the grinding container wall above the blades need to be removed from time to time and returned to the grinding process. The sample is now ready for the detection of amino acids via color reaction with iTAG solution.

Figure 8. Homogenization of bacon; from left to right: original sample; homogenized sample.

4.5. Detection of polychlorinated biphenyls in fish (cutting mill)

The homogenization of fish is a challenge; scales, skin, and bones are fairly resistant to size reduction so that the sample still contains larger pieces after grinding in most mills (e.g., fresh fish in knife mills). A high fat content of the fish makes the process more difficult, as fatty particles stick together to form large lumps which block the mill and keep the sample inhomogeneous. Freeze drying of the fish and further milling in a cutting mill helps to solve the problem. 125 g (four fishes, pre-cut once) of carp or turbot were pulverized in a cutting mill at a revolution speed of 3000 min^{-1}, using a V-rotor which also cuts the scarp and fish bones. The use of a cyclone cools the sample. After 2 min of grinding with a 1-mm bottom sieve, the fish is ground to 1-mm particles without significant heat build-up (**Figure 9**). The sample is now ready for extraction and subsequent gas chromatography.

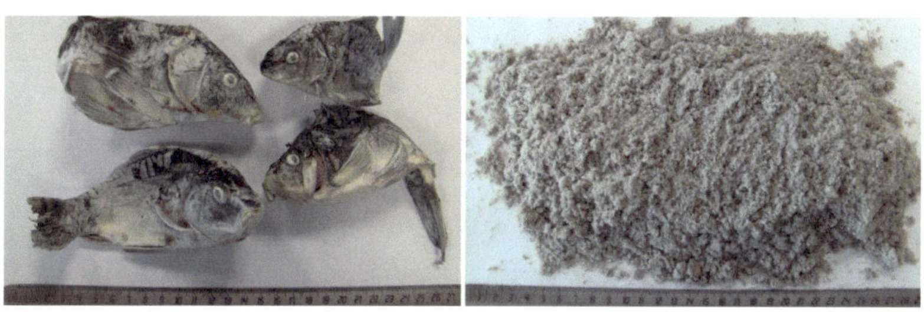

Figure 9. Homogenization of fish; from left to right: original sample, sample ground to <1 mm.

4.6. Pyrrolizidine alkaloids in tea (ultra centrifugal mill)

The group of pyrrolizidine alkaloids comprises 500 chemical compounds which are mostly found in composite flowers, borage family, and leguminous plants. Dried chamomile flowers were processed with the following parameters: a 25-g sample with a maximum particle size of

5 mm was pulverized at a revolution speed of 18,000 min^{-1} in an ultra centrifugal mill using a 0.2-mm ring sieve. After 2 min, the complete sample was ground to a final fineness of <100 μm (**Figure 10**). The use of a cyclone ensures continuous material discharge and cooling of the sample. Thus, the characteristics of the heat-sensitive pyrrolizidine alkaloids are preserved during sample preparation and can be detected by SPE-LC-MS/MS.

Figure 10. Homogenization of chamomile; left to right: original sample, sample ground to <100 μm.

4.7. Ginsenoide in ginseng (mixer mill)

Ginseng has been known for many years in traditional Chinese medicine to have beneficial health effects such as boosting immune reaction and supporting the cardiovascular system. A certain class of chemical substances, such as ginseng saponins, seems to be responsible for the beneficial effects. Therefore, analyzing the composition and content of these substances is of great interest. Small amounts of ginseng roots can be pulverized in mixer mills provided they are smaller than 8 mm. Larger sample pieces must be cut first, for example, by using a cutting mill with a parallel section rotor. 17 mL of pre-cut ginseng particles was pulverized in a mixer mill in a 50-mL stainless steel grinding jar. Fifteen grinding balls with 10-mm diameter were used. After 4 min at a frequency of 30 Hz, a final fineness below 100 μm was achieved (**Figure 11**). The sample was now ready for extraction and subsequent HPLC analysis.

Figure 11. Homogenization of ginseng; left to right: original sample, sample ground to <8 mm, sample ground to <100 μm.

4.8. Mineral determination in large quantities of salt (rotor beater mill)

Rock salt and sea salt not only consist of sodium chloride but may also contain other minerals and silicates, depending on the mining area and method. To analyze the composition of salt, the sample needs to be sufficiently homogenized, considering that larger lumps of rock salt are usually very inhomogeneous. The element concentrations in salt are usually very low so that it is frequently necessary to process amounts in the kilogram range. In principle, a cutting mill could cope with large quantities but the wear would be much greater than in a rotor beater mill, as the cutting bars of the cutting mill are not designed to process large amounts of abrasive materials like salt. With a rotor beater mill, charges of several kilograms can be pulverized easily. Size reduction of the sample is effected by impact and shearing. A distance rotor is used to reduce frictional heat. Thanks to a 5-L collecting vessel, 5 kg of sample with a feed size up to 10 mm is pulverized in one run at a revolution speed of 10,000 min^{-1}. The complete sample is pulverized to less than 200 μm in 6 min (**Figure 12**) and can be analyzed by colorimetric methods or titration.

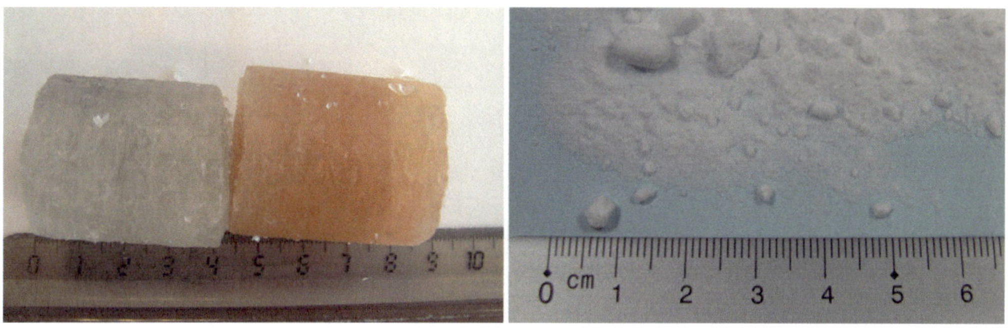

Figure 12. Homogenization of rock salt; left to right: original sample, sample ground to <200 μm.

4.9. Vitamin C analytics in hard candy (knife mill)

Confectionery occurs in very different textures: it can be hard, sticky, greasy, or moist and is frequently inhomogeneous. For HPLC analysis, which is used to detect the content of vitamin C, for example, in hard candy, a particle size distribution between 0.5 and 0.75 mm is ideal. A typical homogenization process in a knife mill involves 100 g of hard candy which is first roughly ground for a few seconds in reverse mode with the blunt side of the knife [8]. The following step involves operation in forward mode with intervals for another 15 s at a revolution speed of 4000 min^{-1}. Further pulverization to a size below 0.5 mm is achieved by grinding for 6–12 s at a revolution speed of 6000 min^{-1} (**Figure 13**). This step-by-step procedure prevents the sample—which has a high sugar and starch syrup content—from sticking to the knife as is often the case in household mixers.

Figure 13. Homogenization of candy; left to right: original sample, sample ground to <500 μm.

4.10. Detection of genetically modified organism in soy beans (knife mill)

Polymerase chain reaction (PCR) is used to detect genetically modified organisms (GMOs) in food. Prior to PCR, the sample must be homogenized. Attention should be paid to sampling and obtaining a representative part sample to ensure meaningful and sensitive GMO testing. From a 20-t bulk of soy beans, a laboratory sample of about 2.5 kg is extracted. For the detection of GMOs a smaller analysis sample, approx. 1000 g in case of corn or soy beans, is extracted from the laboratory sample and thoroughly homogenized in a knife mill. For PCR analysis, only 2 mg of sample material is required. The homogenization step ensures that this 2 mg is a representative of the whole sample. Grainy food like soy beans is processed in a steel container at a revolution speed of 10,000 min^{-1}. With batches of 4 × 250 g, grind sizes below 0.5 mm are obtained within 30 s (**Figure 14**).

Figure 14. Homogenization of soy beans; left to right: original sample, sample ground to <500 μm.

4.11. Further applications: food homogenized at room temperatures

In the following, more application examples for homogenization of food samples at room temperature are given (**Table 2**, **Figure 15**), before putting a focus on cryogenic grinding in the next section.

Sample	Parameters and accessories	Size reduction	Remark
20 g nuts[1]	10,000 min^{-1}, 10 s; grinding container stainless steel	15 to 0.5 mm	High fat content may lead to blockages of sieves in rotor mills
200 g lemons[1]	8000 min^{-1}, 10 s; gravity lid with overflow channels	80 mm to paste	High water content and large particle size: milling only in a knife mill possible
160 g pie [1]	10 s 4000 min^{-1}, 10 s 8000 min^{-1}	30 mm to paste	Starting with short intervals at the set speed helps to avoid material sticking on walls of grinding container
280 g lasagna[1]	10 s 4000 min^{-1}, 20 s 8000 min^{-1}	80 mm to paste	
500 g bread[1]	1 min 4000 min^{-1}, knife with titanium-niob coated blades	160 to 1.5 mm	Heavy metal determination: knife with titanium-niob coated blades was used
100 g dried pear[1]	15 s 4000 min^{-1}, 15 s 7000 min^{-1}	50 to 1 mm	Homogenization of sticky material
200 g hard cheese[1]	10 s 2000 min^{-1}, 10 s 6000 min^{-1}	20 to 1.5 mm	Finer particles could be obtained under cryogenic conditions

Sample	Parameters and accessories	Size reduction	Remark
800 g soup[1]	30 s 4000 min^{-1}, with interval	50 mm to paste	Double-sealed lid for liquid samples, interval mode improves sample mixing
700 g pizza[1]	90 s 2000 min^{-1}, with interval	50 to 3 mm	Cryogenic grinding can achieve further homogenization
150 g ginger[1]	35 s 4000 min^{-1} reverse mode	30 to 0.8 mm	Reverse mode helps to avoid wear of blades when cutting tough material
5 eggs[1]	10 s 10000 min^{-1}	70 mm to paste	Very fast homogenization
100 g field bean[2]	12 tooth rotor, distance sieve 1 mm, 60 s, 18000 min^{-1}	15 to 0.5 mm	To avoid warming, the sample is filled into the mill slowly but continuously. The distance sieve is used to reduce heat.
150 g gelatin[2]	12 tooth rotor, distance sieve 1 and 0.35 mm, 45 and 120 s, 18000 min^{-1}, cyclone	70 to 0.5 mm	Distance sieve to avoid warming, slow feeding required, cyclone helps to cool sample and improve sample discharge
1200 g salt[2]	12 tooth rotor, distance sieve 0.08 mm, 10 min, 18000 min^{-1}, cyclone and vibratory feeder	1 mm to 15 μm	Sample is hygroscopic and may stick; check sample discharge from time to time, vibratory feeder facilitates feeding of larger quantities
50 g green coffee[2]	12 tooth rotor, distance sieve 0.75 mm, 3 min, 18000 min^{-1}, cyclone	15 to 0.75 mm	Distance sieve and cyclone reduce heat and fat release. Sieves with small aperture sizes may be blocked due to fat release.
150 g corncob[3]	Parallel section rotor, bottom sieve 4 mm, 1500 min^{-1}, 20 s; & 12 tooth rotor, ring sieve 0.5 mm, 20 s, 18,000 min^{-1}	150 mm to 400 μm	Grinding in two steps as initial sample is too large for direct feeding into ultra centrifugal mill; required final fineness achieved efficiently in ultra centrifugal mill
50 g viola roots[3]	Six-disc rotor, bottom sieve 4 mm, 1500 min^{-1}, 20 s & 12 tooth rotor, ring sieve 0.5 mm, 15 s, 18,000 min^{-1}	100 mm to 200 μm	Sample is too hard for manual pre-crushing, fine grinding in ultra centrifugal mill as second step yields very fine material
5 kg tea[4]	V-rotor, 0.25 mm bottom sieve, 3000 min^{-1}, 25 min	6 cm to 200 μm	Less warming of the sample compared to ultra centrifugal mill but same fineness and time
10 pieces, gelatin blocks[4]	V-rotor, 6 mm bottom sieve, 3000 min^{-1}, 10 s, cyclone	8 × 5 × 1 cm to 6 mm	The cyclone is used to increase sample discharge from grinding chamber (very light material)
10 kg oat[4]	Parallel section rotor, 6 mm bottom sieve, 700 min^{-1}, 60 s, cyclone	6 to 3 mm	Reduction of speed increases obtained particles size, fine fraction is reduced
50 g mushrooms[4]	Parallel section rotor, 6 mm bottom sieve, 1500 min^{-1}, 10 s	30 to 4 mm	Sample was ground piece by piece, high-coarse particle content required
2 l manioc[5]	0.25 mm 360° sieve, cyclone, feeder, 10,000 min^{-1}, 11 min	2 mm to 200 μm	Vibratory feeder for larger quantities
20 kg roasted milk with sugar[5]	2 mm 360° sieve, cyclone and feeder, 30 l receptacle, 10,000 min^{-1}, 38 min	3 to 1 mm	Distance sieve reduces sticking of sample; 30 l receptacle required for large sample quantity

Sample	Parameters and accessories	Size reduction	Remark
2 kg herbs[5]	0.08 mm 360° sieve, cyclone, feeder, 30 l receptacle, 10000 min^{-1}, 80 min	15 mm to 120 µm	Vibratory feeder for large quantities, 80 min to process this large sample quantity to a fineness of 120 µm
30 g corn[6]	0.5 mm sieve, 14,000 min^{-1}, 15 s	10 to 0.3 mm	Quick and contamination-free grinding of non-fatty samples, high-sample throughput
100 g barley[6]	1 mm sieve, 14,000 min^{-1}, 10 s	10 to 1 mm	
50 g dry noodles[6]	2 mm sieve, 14,000 min^{-1}, 20 s	15 to 0.75 mm	

[1]Knife mill.
[2]Ultra centrifugal mill.
[3]Cutting mill and ultra centrifugal mill.
[4]Cutting mill.
[5]Rotor beater mill.
[6]Cyclone mill.

Table 2. Application examples of food homogenized at room temperature.

Figure 15. Food samples which can be pulverized at room temperature; first row from left to right: pistachios, lemon, pie, lasagna, bread, pears, cheese, soup; second row from left to right: ginger, beans, green coffee, corncob, viola root, maize, barley, noodles; third row from left to right: manioc, roasted milk and sugar, herbs, block of gelatin, salt, pizza, nuts, sausages.

5. Special application: cryogenic grinding of food samples

Most sample materials can be ground to the required analytical fineness at room temperature. However, there are limits, for example, when even a small temperature increase affects the

http://dx.doi.org/10.5772/intechopen.68811

sample in a negative way, or when the material is very elastic and will only be deformed. Moreover, food samples, which are fatty or sticky, may block the mill. Cryogenic grinding is the best way to pulverize food samples when they are sticky, fatty, semi-liquid samples (e.g., cheese, raisins, wine gum, or marzipan), and simply clump together when ground at room temperature. In a cryogenic-grinding process, the samples don't clump and are effectively homogenized. Under cryogenic conditions, the loss of volatile ingredients like alcohol can be limited or residues of softeners, which migrate from plastic wrappings into fatty food like meat, are preserved. Such ingredients would escape when the sample is warmed during grinding. Furthermore, cold milling preserves the original structures of vitamins or proteins. Cryogenic grinding is carried out with grinding aids such as liquid nitrogen LN_2 (−196°C) or dry ice (solid CO_2; −78°C) which embrittle the sample and make it break more easily. In this section, the special requirements for cryogenic grinding in different mills will be discussed as well as which other aspects need to be taken into consideration (**Table 3**). Basically, all rules and recommendations described for grinding at room temperature must be observed for cryogenic grinding, too.

5.1. Cryogenic grinding in mixer mills

It is important to fill the jar first with the grinding ball(s) and with the sample and close it tightly before embrittling. Care must be taken that no LN_2 is enclosed in the grinding jars because the evaporation of the LN_2 would result in a considerable pressure increase inside the grinding jar. The closed grinding jars, and thus the sample, are embrittled in a LN_2 bath for 2–3 min. Suitable grinding jars for cryogenic grinding are made of steel or PTFE; it is not recommended to use jars made of different materials (e.g., steel jar with lining of zirconium oxide). This is important, as two different materials may react differently to extreme temperatures of −196°C which may lead to damages of the jar. Single-use vials of 1.5, 2, and 5 mL are also available for cryogenic grinding. Due to the high-energy input and the resulting frictional heat, the grinding process should not take longer than 2 min to prevent the sample from warming up and to preserve its breaking properties. If longer grinding times are required, these should be interrupted by intermediate cooling of the closed grinding jars.

5.2. Cryogenic grinding in cryo mills

Cryo mills offer the advantage of continuous cooling of the grinding jar with LN_2, reducing the temperature of jar and sample to −196°C within minutes. Thus, a consistent temperature of −196°C is guaranteed even for long grinding times without the need for intermediate cooling breaks. Moreover, care should be taken that the user comes at no point into contact with LN_2. An automatic pre-cooling function ensures that the grinding process does not start before a temperature of −196°C is reached and maintained. For heavy-metal-free grinding, a zirconium oxide grinding jar should be used. Further suitable materials are stainless steel or single-use vials (1.5 or 2 mL). Just like in mixer mills, embrittlement of the sample occurs indirectly as the sample is enclosed in the grinding jar.

Mill	Feed size and max feed quantity (both depending on sample material)	Remark
Mixer mill	<8 mm 2 × 20 mL	• Sample is placed in leak-free grinding jar of steel or PTFE and embrittled before grinding, LN_2 preferred over dry ice • Intermediate cooling may be required
Cryo mill	<8 mm 1 × 20 mL	• Continuous grinding at −196°C with LN_2 • User comes at no point in contact with LN_2 • Zirconium oxide grinding jar available for cryogenic grinding
Ultra centrifugal mill	<10 mm 4000 mL	• Embrittlement with dry ice or LN_2 • Dry ice preferred if sample material is <1 mm or has low thermal capacity • Use of cyclone mandatory
Knife mill	<40 mm 2000 mL	• Embrittlement with dry ice • Dry ice cools sample during grinding • Use of full metal knife, grinding container of stainless steel and specific lid mandatory
Cutting mill	<80 mm 4000 mL	• Cryogenic grinding with dry ice or LN_2 • Use of six-disc rotor and cyclone mandatory • Bottom sieves 2–20 mm suitable

Table 3. Overview of mills suitable for cryogenic grinding.

5.3. Cryogenic grinding in ultra centrifugal mills

Ultra centrifugal mills accept larger sample volumes than mixer mills. The sample is directly immersed into a container filled with LN_2 before being continuously but slowly fed to the hopper of the mill with a steel spoon. When using dry ice as grinding aid, this needs to be mixed with the sample (one part sample, two parts dry ice) and the entire mixture is then pulverized in the mill. Using a cassette in combination with a cyclone is recommended for cryogenic grinding to ensure that the evaporating cooling agent is completely discharged during the grinding process. The use of dry ice rather than LN_2 should be preferred if the sample is already smaller than 1 mm, as the transfer of a dry ice-sample mixture to the mill is much easier than fishing the sample with a spoon from the LN_2 bath. Also, if the sample has a low thermal capacity, dry ice is also preferable as it cools the sample during grinding.

5.4. Cryogenic grinding in knife mills

Sticky and tough food samples such as cheese, raisins, wine gum, or marzipan are perfectly homogenized in a knife mill. The use of LN_2 is not recommended as the knife mills are not designed for temperatures as low as −196°C. Even chocolate, which simply becomes paste-like when processed at room temperature, can be successfully pulverized cryogenically. The sample is mixed with dry ice in a ratio of 1:2; after a few minutes, it is thoroughly cooled and the grinding process starts. The dry ice keeps the sample cool all the time. Care should be taken not to use any plastic accessories when carrying out cryogenic grinding in the knife mills as these could be damaged during the process. Suitable accessories include a grinding container of stainless steel, a full metal knife, and a lid with aperture to allow evaporation of the gaseous carbon dioxide.

5.5. Cryogenic grinding in cutting mills

Cutting mills are particularly suitable for processing larger feed sizes than ultra centrifugal mills or knife mills. Both the use of LN_2 and dry ice are possible (see Section "Cryogenic grinding with ultra centrifugal mills" for advantages of using dry ice). The embrittled sample material is rather hard; therefore, the use of the six-disc rotor is recommended as it works more like a shredder. It is also suitable to cut heterogeneous samples such as frozen chicken parts including bones.

5.6. Cryogenic applications: food homogenized at low temperatures

Table 4 provides an overview of samples which are best ground under cryogenic conditions (**Figure 16**).

Sample	Parameters and accessories	Size reduction	Remark
10 jelly bears[1]	1 min, 30 Hz	20 mm to 300 μm	Tongs to transfer the grinding jars from LN_2 bath into mill. Grinding in 50 mL with grinding ball 25 mm grinding jar (both stainless steel)
20 g chewing gum[1]	30 s, 30 Hz	15 mm to 500 μm	
10 g liver[1]	2 min, 30 Hz	6 mm to 400 μm	
10 g cookies[1]	1 min, 30 Hz	12 mm to 300 μm	
3 g vanilla pod[1]	20 s, 30 Hz	10 mm to 500 μm	
2 g cherries[2]	10 s, 30 Hz	15 mm to 600 μm	Food samples: pre-cooling of appr. 5 min is typical. Usually grinding is done in 50-mL grinding jar with grinding ball 25 mm (both stainless steel)
8 g pork[2]	3 min, 30 Hz	10 mm to 200 μm	
6 g licorice[2]	2 min, 30 Hz	10 mm to 300 μm	
9 g green coffee[2]	15 min, 30 Hz	10 mm to 150 μm	
5 g cheese[2]	2 min, 30 Hz	8 mm to 300 μm	
1 praline; liquid filling[2]	2 min, 30 Hz	10 mm to 400 μm	

Sample	Parameters and accessories	Size reduction	Remark
500 g wine gum[3]	40 s 2000 min^{-1} reverse; 20 s 4000 min^{-1} forward	20 to 0.8 mm	Grinding container stainless steel, full metal knife, cryo lid with aperture; dry ice. Pre-cutting in reverse mode reduces wear of blades. Intervals can help to improve sample mixing.
250 g grapes[3]	15 s 2000 min^{-1} reverse; 15 s 4000 min^{-1} forward	20 mm to 400 μm	
300 g block of marzipan[3]	20 s 2000 min^{-1} reverse; 20 s 4000 min^{-1} forward	40 mm to 800 μm	
400 g pure bacon[3]	45 s 2000 min^{-1} reverse; 30 s 4000 min^{-1} forward	30 to 1 mm	
800 g raisin[3]	45 s 2000 min^{-1} reverse	15 to 0.5 mm	
100 g cereals[4]	12 tooth rotor, ring sieve 0.5 mm, 3 min, 18000 min^{-1}	8 mm to 250 μm	Use of cyclone and LN_2
70 g nutritionals[4]	12 tooth rotor, ring sieve 0.12 mm, 5 min, 18,000 min^{-1}	2 mm to 100 μm	Use of cyclone and dry ice. Grinding in two steps using two different ring sieves is efficient if initial sample size is larger.
100 g dried apples[4]	12 tooth rotor, ring sieve 0.5 mm, 1 min, 18000 min^{-1}	5 mm to 250 μm	
15 g toffee candy[4]	12 tooth rotor, ring sieve 2 mm and 0.5 mm, 1 min, 18,000 min^{-1}	10 mm to 500 μm	
500 g block of chocolate[5]	Parallel section rotor, 4 mm bottom sieve, 700 min^{-1}, 60 s	40 to 4 mm	Use of cyclone and LN_2; reduced speed leads to less heat build-up
1 kg trout[5]	6-disc rotor, 20 mm bottom sieve, 700 min^{-1}, 60 s	200 to 20 mm	
500 g lump of cocoa butter[5]	six-disc rotor, 6 mm bottom sieve, 700 min^{-1}, 90 s	100 to 6 mm	
20 kg sweet potatoes[5]	6-disc rotor, 20 mm bottom sieve, 1500 min^{-1}, 15 min	100 to 20 mm	Use of cyclone and dry ice

[1]Mixer mill.

[2]Cryo mill.

[3]Knife mill.

[4]Ultra centrifugal mill.

[5]Cutting mill.

Table 4. Application examples of cryogenically homogenized food.

Figure 16. Food samples ground under cryogenic conditions; first row from left to right: jelly bears, chewing gum, fresh liver, cookies, vanilla pods, dried cherries, fresh meat; second row from left to right: licorice, cheese, praline with liquid core, wine gum, grapes, block of marzipan, bacon; third row from left to right: raisins, onions, bar of cereals, dried apples, block of chocolate (pre-ground), frozen fish (pre-ground), cocoa butter (pre-ground).

6. Conclusion

In this chapter, it was demonstrated by a wealth of application examples that sample preparation prior to any food analysis is an essential step of the quality control process as only fully homogenized samples provide reliable and reproducible analysis results. Due to the wide range of laboratory mills and accessories available, it is important to consider all aspects of the sample preparation process before selecting a suitable device to make this important step prior to sample analysis most efficient and reliable. Both the knowledge of the sample characteristics and the available types of mills and accessories enable the user to process these samples with a minimum of time and effort but with best possible results.

7. General remark

See more detailed information on our webpage www.retsch.com—the different application reports, brochures, "the sample 43," and the "Art of Milling" may be downloaded.

Author details

Tanja Butt

Address all correspondence to: t.butt@retsch.com

Retsch GmbH, Haan, Germany

References

[1] Hanke T. Komplexe Proben schnell und reproduzierbar zerkleinern. DLR. 2014;**4**:189-192

[2] Theisen A. As fine as necessary. Chemical Plants+ Processing. 2012;**2**

[3] Retsch. Expert Guide Milling [Internet]. 2016. Available from: http://www.retsch.com/downloads/brochures/

[4] Retsch. White Paper: Cryogenic Preparation of Sample Materials [Internet]. 2016. Available from: http://www.retsch.com/downloads/application-reports/

[5] Theisen A. Grain & feed milling technology. 2011-October;(18-20)

[6] Theisen A. Detection of mycotoxins in nuts. Labasia. 2009;**16/6**:33-34

[7] Retsch. Customer Magazine. Food Analysis—Reproducible Sample Preparation with RETSCH Laboratory Mills [Internet]. 2016. Available from: http://www.retsch.com/downloads/customer-magazine-the-sample/dl_details/1/file/55acca6a-6ab4-49de-b953-521dbc282b87/

[8] Hanke T. Sample preparation for HPLC analysis of confectionery. International Labmate. 2015-February;(24-25)

Chapter 3

Bioaerosols in the Food and Beverage Industry

Shirleen M. Theisinger and Olga de Smidt

Additional information is available at the end of the chapter

http://dx.doi.org/10.5772/intechopen.69978

Abstract

Bioaerosol monitoring is a rapidly emerging area of industrial hygiene. Microbial roles in atmospheric processes are thought to be species specific and potentially depend on cell viability. Accumulating evidence suggests that exposure to bioaerosols may cause adverse health effects, including disease. Studies of bioaerosols have primarily focused on chemical composition and biological composition, and the negative effects thereof on ecosystems and human health have largely gone unnoticed. This gap can be attributed to international standards on acceptable maximum bioaerosol loads not being uniform and the lack of uniform standardized methods for collection and analysis of bacterial and fungal bioaerosols. In this chapter, bioaerosol composition, relevance of bioaerosols to the food processing facility, sampling and detection approaches, and complications were discussed.

Keywords: bioaerosols, microbial diversity, passive/active sampling, food handler health

1. Introduction

Microbes are ubiquitous in the environment and play key functional roles in nearly all ecosystems [1]. Indeed, environmental bacteria, fungi and viruses are a part of our natural environment, having coevolved with all the other living organisms, including humans. Airborne dissemination is a natural and necessary part of the life cycle of many microbes [2]. Bioaerosols originate from all types of environments, including atmosphere, soil, freshwater and oceans, and their dispersal into air is temporally and spatially variable. Bioaerosols are emerging as important, yet poorly understood players in atmospheric processes. Research on bioaerosols has experienced and continues to experience stellar growth [3].

In 1861, the first measurements of airborne microbes were reported by Louis Pasteur in the Journal Annales des Sciences Naturelles [4]. A century later, research into the role of bioaerosols in occupation-related diseases mainly focuses on noninfectious diseases. Pepys and coworkers [5] first demonstrated that patients with existing diseases are more likely to suffer attacks of farmer's lung when inhaling spores from thermophilic actinomcetes. Byssinosis among cotton workers was an important research topic during the 1970–1980s. The most likely causative agents for this disease were Gram-negative bacteria, and the endotoxins located in their outer cell wall [6]. The interest in bioaerosol exposure has increased over the last few decades, largely born from the direct association of bioaerosols with a wide range of adverse health effects. These effects can have major public health impacts which include contagious infectious diseases, acute toxic effects, allergies and cancer [7]. Furthermore, bioaerosols could potentially settle on surfaces and equipment and contribute to safety or spoilage risks where food is prepared, processed or packaged [8].

Due to the presence of great amounts of organic matter, the release of bioaerosols can be very high in certain industrial sectors such as agriculture, all types of food industries, waste management facilities, textile and wood industries. Each bioaerosol sample is unique as its composition varies in time and space (abundance and diversity of species, quantity of pro-inflammatory components). This often leads not only to high variation between samples from the same workplace, which can be due to external factors but also to the dynamic evolution of the colonized substrate and the fast multiplication rate of many microbes.

In this chapter, bioaerosol composition, relevance of bioaerosols to the food processing facility, approaches and complications in detection and approaches to sampling bioaerosols will be discussed.

2. Bioaerosols composition

An aerosol is a two-phase system of gaseous phase (air) and particulate matter (dust, pathogens), thus making it an important microbial vehicle. Bioaerosols are defined as "aerosols comprising of particles of biological origin or activity which may affect living things through infectivity, allergenicity, toxicity, pharmacological or other processes" [9, 10]. Bioaerosols are a diverse collection of small pieces of material emitted directly from the biosphere into atmosphere [11].

Bioaerosols are globally ever present, in some cases can dominate suspended particle concentrations and comprise a diverse selection of particle types, including whole organisms (bacteria, mold, fungi, yeast and algae), reproductive entities (pollen, spores from fungi, bacteria, ferns and mosses), biopolymers (DNA, chitin, cellulose and other polysaccharides), plant debris, insect parts, and decaying biomass. The components of bioaerosols range in size; pollens from anemophilous plants have typical diameters of 17–58 μm, fungal spores are typically 1–30 μm in diameter, bacteria are typically 0.25–8 μm in diameter, and viruses are typically less than 0.3 μm in diameter. Furthermore, fragments of plants and animals may vary in size. Apart from the fact that bioaerosol particles can span several orders of magnitude in diameter, bacteria may also occur as clusters of cells or may be dispersed into the air on plants or animal fragments, on soil particles, on pollen or on spores that have become airborne [12]. All these characteristics contribute to making accurate analysis of bioaerosols very challenging.

2.1. Microbial component

Microbes are ubiquitous in nature and also present in the air as living cells able to infect or contaminate the surface or tissue it settles in or upon. These airborne bacterial and fungal cells can reach concentrations of 10^3 and 10^5 cells m^{-3}, respectively [7]. **Table 1** lists different bacterial, yeast and mold genera detected as bioaerosol components found in food industries from noteworthy research since 2003. The table depicts only data from food-related industries where microbial components were detected and identified to at least genus level. Research focused on viability testing only (total plate counts, total yeast and mold) was not mentioned.

Genus	Occupational environment	Sampling method (sampler)
Bacteria		
Acinetobacter	Milk processing [13] Abattoir (beef) [8]	Active: impaction (MAS-100) Passive: petri plate
Arthrobacter	Milk processing [13]	Active: impaction (MAS-100)
Bacillus	Milk processing [13] Food warehouse (rice grains) [14] Abattoir (beef/pork) [8]	Active: impaction (MAS-100) Passive: petri plate Passive: petri plate
Brevibacterium	Milk processing [13] Abattoir (pork) [8]	Active: impaction (MAS-100) Passive: petri plate
Brevundimonas	Milk processing [13]	Active: impaction (MAS-100)
Brochothrix	Abattoir (pork) [8]	Passive: petri plate
Cedecea	Abattoir (beef/pork) [8]	Passive: petri plate
Cellulomonas	Abattoir (pork) [8]	Passive: petri plate
Chryseobacterium	Milk processing [13] Abattoir (pork) [8]	Active: impaction (MAS-100) Passive: petri plate
Chryseomonas	Abattoir (beef/pork) [8]	Passive: petri plate
Citrobacter	Abattoir (beef) [8]	Passive: petri plate
Curtobacterium	Milk processing [13]	Active: impaction (MAS-100)
Enterobacter	Food warehouse (rice grains) [14] Abattoir (beef/pork) [8]	Passive: petri plate Passive: petri plate
Escherichia	Food warehouse (rice grains) [14] Abattoir (beef/pork) [8]	Passive: petri plate Passive: petri plate
Flavimonas	Abattoir (beef/pork) [8]	Passive: petri plate
Frigoribacterium	Milk processing [13]	Active: impaction (MAS-100)
Klebsiella	Abattoir (pork) [8]	Passive: petri plate

Genus	Occupational environment	Sampling method (sampler)
Kluyvera	Abattoir (beef/pork) [8]	Passive: petri plate
Kocuria	Milk processing [13]	Active: impaction (MAS-100)
	Abattoir (beef/pork) [8]	Passive: petri plate
Leclercia	Abattoir (pork) [8]	Passive: petri plate
Leuconostoc	Milk processing [13]	Active: impaction (MAS-100)
Lysinibacillus	Milk processing [13]	Active: impaction (MAS-100)
Macrococcus	Milk processing [13]	Active: impaction (MAS-100)
Massilia	Milk processing [13]	Active: impaction (MAS-100)
Micrococcus	Noodle manufacturing [15]	Active: impaction (MAS-100)
	Milk processing [13]	Active: impaction (MAS-100)
	Abattoir (beef/pork) [8]	Passive: petri plate
Microbacterium	Milk processing [13]	Active: impaction (MAS-100)
	Abattoir (beef/pork) [8]	Passive: petri plate
Moraxella	Milk processing [13]	Active: impaction (MAS-100)
	Abattoir (beef) [8]	Passive: petri plate
Morganella	Abattoir beef/pork) [8]	Passive: petri plate
Nesterenkonia	Abattoir (beef/pork) [8]	Passive: petri plate
Novosphingobium	Milk processing [13]	Active: impaction (MAS-100)
Paenibacillus	Abattoir (beef) [8]	Passive: petri plate
Pantoea	Abattoir (beef/pork) [8]	Passive: petri plate
Pedobacter	Milk processing [13]	Active: impaction (MAS-100)
Proteus	Food warehouse (rice grains) [14]	Passive: petri plate
Pseudomonas	Milk processing [13]	Active: impaction (MAS-100)
	Food warehouse (rice grains) [14]	Passive: petri plate
	Abattoir (beef/pork) [8]	Passive: petri plate
Rahnella	Milk processing [13]	Active: impaction (MAS-100)
Rhodococcus	Milk processing [13]	Active: impaction (MAS-100)
Roseomonas	Milk processing [13]	Active: impaction (MAS-100)
Salmonella	Abattoir (beef/pork) [8]	Passive: petri plate
Serratia	Abattoir (pork) [8]	Passive: petri plate
Shigella	Abattoir (beef/pork) [8]	Passive: petri plate
Spingomonas	Milk processing [13]	Active: impaction (MAS-100)
Staphylococcus	Noodle manufacturing [15]	Active: impaction (MAS-100)
	Milk processing [13]	Active: impaction (MAS-100)
	Broiler chicken barn [16]	Active: impaction (MAS-100)
	Food warehouse (rice grains) [14]	Passive: petri plate
	Abattoir (beef/pork) [8]	Passive: petri plate

Genus	Occupational environment	Sampling method (sampler)
Stenotrophomonas	Abattoir (beef/pork) [8]	Passive: petri plate
Streptococcus	Food warehouse (rice grains) [14]	Passive: petri plate
Wautersiella	Milk processing [13]	Active: impaction (MAS-100)
Yeast		
Candida	Fruit juice production (unpublished data)	Active: impaction (SAMPL'AIR™)
Cryptococcus	Fruit juice production (unpublished data)	Active: impaction (SAMPL'AIR™)
Meyerozyma	Fruit juice production (unpublished data)	Active: impaction (SAMPL'AIR™)
Pichia	Fruit juice production (unpublished data)	Active: impaction (SAMPL'AIR™)
Rhodotorula	Fruit juice production (unpublished data)	Active: impaction (SAMPL'AIR™)
Wickerhamomyces	Fruit juice production (unpublished data)	Active: impaction (SAMPL'AIR™)
Molds		
Absida	Wheat flour mill [17]	Active: impaction (RCS)
Alternaria	Food warehouse (rice grains) [14] Wheat flour mill [17] Rice mill [18] Cake factory [19]	Passive: petri plate Active: impaction (RCS) Active: impaction (six-stage viable Andersen cascade) Passive: petri plate
Aspergillus	Noodle manufacturing [15] Food warehouse (rice grains) [14] Wheat flour mill [17] Rice mill [18] Cake factory [19]	Active: impaction (MAS-100) Passive: petri plate Active: impaction (RCS) Active: impaction (six-stage viable Andersen cascade) Passive: petri plate
Aureobasidium	Rice mill [18]	Active: impaction (six-stage viable Andersen cascade)
Botrytis	Food warehouse (rice grains) [14]	Passive: petri plate
Cephalosporium	Food warehouse (rice grains) [14]	Passive: petri plate
Cercospora	Food warehouse (rice grains) [14]	Passive: petri plate
Cladosporium	Noodle manufacturing [15] Food warehouse (rice grains) [14] Wheat flour mill [17] Rice mill [18] Cake factory [19]	Active: impaction (MAS-100) Passive: petri plate Active: impaction (RCS) Active: impaction (six-stage viable Andersen cascade) Passive: petri plate
Colletotrichum	Food warehouse (rice grains) [14]	Passive: petri plate
Curvularia	Food warehouse (rice grains) [14]	Passive: petri plate
Epicoccum	Wheat flour mill [17] Rice mill [18]	Active: impaction (RCS) Passive: petri plate
Eurotium	Wheat flour mill [17]	Active: impaction (RCS)

Genus	Occupational environment	Sampling method (sampler)
Fusarium	Food warehouse (rice grains) [14]	Passive: petri plate
	Wheat flour mill [17]	Active: impaction (RCS)
Helminthosporium	Food warehouse (rice grains) [14]	Passive: petri plate
Mortierella	Food warehouse (rice grains) [14]	Passive: petri plate
Mucor	Food warehouse (rice grains) [14]	Passive: petri plate
	Wheat flour mill [17]	Active: impaction (RCS)
	Rice mill [18]	Active: impaction (six-stage viable Andersen cascade)
	Cake factory [19]	Passive: petri plate
Penicillium	Noodle manufacturing [15]	Active: impaction (MAS-100)
	Food warehouse (rice grains) [14]	Passive: petri plate
	Wheat flour mill [17]	Active: impaction (RCS)
	Rice mill [18]	Active: impaction (six-stage viable Andersen cascade)
	Cake factory [19]	Passive: petri plate
Rhizopus	Food warehouse (rice grains) [14]	Passive: petri plate
	Wheat flour mill [17]	Active: impaction (RCS)
	Rice mill [18]	Active: impaction (six-stage viable Andersen cascade)
	Cake factory [19]	Passive: petri plate
Stachybotrys	Food warehouse (rice grains) [14]	Passive: petri plate
Trichoderma	Food warehouse (rice grains) [14]	Passive: petri plate
	Cake factory [19]	Passive: petri plate
Verticillium	Food warehouse (rice grains) [14]	Passive: petri plate

Table 1. Different microbial genera detected as bioaerosol components in food production, processing and storage environments.

Despite the wide diversity detected, not all have been directly indicated as spoilers or contaminants of food or of being the causative agents of disease due to bioaerosol exposure. Furthermore, not all species in a genus are necessarily harmful, which emphasizes using the appropriate sampling technique and identification methods to suite the objective for bioaerosol testing. Although all microbes present in the air may not be harmful as pathogens in vegetative state, their spores, toxins, endospores, LPS and other constituents have been linked to disease and could pose risk.

2.1.1. Spores

Bioaerosols contain mostly spores that are tougher, metabolically less active and often better adapted to dispersal. Spores are single or multicellular units surrounded by a rigid cell wall. Each spore is capable of reproducing the entire organism.

Certain bacteria can survive adverse environmental conditions for prolonged periods by producing a thick-walled spore structure called an endospore. Endospores function to protect the bacterial DNA against the conditions or substances in the environment that would lead to the destruction of nonendospore-forming bacteria [20]. *Bacillus cereus* is one such spore-forming bacterium that naturally occurs in many foods. *B. cereus* form spores that are resistant to heating and dehydration, and when food-containing *B. cereus* spores are in the "temperature danger zone," the spores geminate and the bacteria grow and produce toxins that cause illness in humans. *B. cereus* can cause vomiting or diarrhea, and, in some cases, both depend on the kinds of toxin it produces [21].

Mold spores are somewhat resistant to destruction, and they are not usually pathogenic to humans. Epidemiological and experimental studies support the fact that *Aspergillus* spp. are highly allergenic molds. These molds are known to cause two allergic diseases of the respiratory system: bronchial asthma and allergic rhinitis. Spore concentrations of above 50 CFU m^{-3} have been associated with higher prevalence of sick-building syndrome [22, 23].

2.1.2. Toxins

Endotoxins are composed of lipopolysaccharides and lipooligosaccharides associated with proteins and lipids and are part of the exterior cell membrane of Gram-negative bacteria. Endotoxins are either present in the fragments of the cell wall or in the bacterial cell released during bacterial lysis. Endotoxins are nonallergenic, with strong pro-inflammatory properties. They are present in many occupational environments: ambient air and house dust [24]. Induction of airway inflammation and dysfunction can be attributed to the inhalation of endotoxins [25]. Endotoxin exposure has been associated with the occurrence of respiratory disorders, including asthma-like symptoms, chronic airway obstruction, bronchitis, increased airway responsiveness, and byssinosis [26]. Unlike molds, endotoxin has also been recognized as a causative factor in the ethnology of occupational lung diseases, including nonallergic asthma and organic dust toxic syndromes [27, 28].

During the nutrient degradation process, fungi release secondary metabolites called mycotoxins. Mycotoxins are toxic fungal metabolites produced by molds in vegetal matrices and could be potentially detected in bioaerosols because of their adsorption on spores and dust particles [29, 30]. Mycotoxins are nonvolatile compounds and will be found in the air only if the environment in which they are produced is disturbed. These molecules act as defense mechanism against other microbes, including other fungi. A given fungal species may produce different toxins depending on the substrate and local environmental factors. Mycotoxins and their associated health effects through respiratory exposure are not well known. They could be the causal agents of effects reported following exposure to molds. Reported symptoms include skin and mucous membrane irritation, nausea, headache, immunosuppression and systemic effects such as dizziness and cognitive and neuropsychological effects [22, 31, 32].

2.1.3. Other

Other bioaerosol components of microbial origin considered nonviable, but bioactive may be present in the air. β-(1-3)-D-glucan is a polymer glucose of high molecular weight found in the cell walls of bacteria, molds, and plants [31]. They consist of glucose polymers with variable molecular weight and degree of branching [24]. β-(1-3)-D-glucan is associated with dry cough, cough associated with phlegm, hoarseness and atopy and has been reported in indoor environments [33]. Part of the components of the cell wall of Gram-positive bacteria consists of peptidoglycans. With the inhalation of Gram-positive bacteria, these peptidoglycans may be potential casual agents of lung inflammation [31].

During bacterial growth or cell death, proteins are normally secreted that are bioactive molecules called exotoxins. Exotoxins are usually associated with infectious diseases such as cholera, tetanus and botulism, but they can also be found on surfaces that can take on an aerosol form and could support bacterial growth [31].

3. Relevance to the food processing facility

Airborne particles and bioaerosols are easily transported, transferred and displaced from one environment to the other. Complex mixtures of bioaerosols such as fungi, allergens, and bacteria along with nonbiological particles (e.g., dust, smoke, particles generated by cooking, organic, and inorganic gases) are contained in indoor environments [34]. The bioaerosols and their components could pose an environmental hazard when presented in high concentrations in indoor environments, resulting in spoilage/contamination of food products or occupational health risks [35].

3.1. Food product–related risk: spoilage or contamination

Before spoilage becomes obvious, microbes have begun the process of breaking down food molecules for their own metabolic needs, resulting in a variety of sensory cues such as off-colors, off-odors, softening of fruits and slime. Firstly, the sugars are easily digested carbohydrates, then plant pectins are degraded, and proteins are attacked and produce volatile compounds with characteristic smells such as amines, ammonia, and sulfides. Early detection of spoilage would be advantageous in reducing food loss because there may be interventions that could halt or delay deterioration. Several methods for determining concentrations of spoilage microbes or volatile compounds produced by spoilage microbes have been devised. Many of these methods are considered insufficient as they are time consuming and/or do not give constant, reliable results and are labor intensive [31].

Food can also be contaminated by the presence of harmful chemicals and microbes which can cause illness when consumed. For this reason, traceability and source determination of contamination remain a relevant topic in food preservation research [36]. Bioaerosols implicated in respiratory-associated hazards have received much attention, but the potential of food-associated microbes and food-borne pathogens in bioaerosols to cause food spoilage needs to be clarified. Evidence exists that pathogenic microbes are found in the air, and that these microbes are present in certain products. However, traceable evidence of bioaerosols as the causative agent of spoilage or contamination of food products is not readily available.

3.2. Food handler-related risk: occupational health

Exposure to higher risks of biological hazards is characteristic to certain industries such as health care, agriculture, fishery, some food industries, construction, and mining. Workers employed in these industries have higher prevalence of respiratory diseases and airway inflammation [37]. It is difficult to conduct a comprehensive evaluation of personal bioaerosol exposure in occupational or indoor environments [38], owed to the complex composition of bioaerosols, and the lack of standardized sampling/analysis methods [37]. Without appropriate personal exposure assessment and standardized sampling/analysis methods, establishing dose relationships and relevant exposure guidelines are difficult.

Exposure to bioaerosols in the occupational environment is associated with a wide range of health effects including infectious diseases, acute toxic effects, allergies, and cancer. These possibilities have been studied for the last 20 years; several cases of pulmonary cancers were reported in workers exposed to aflatoxins via respiratory route [39, 40]. In Denmark, an increase in the risk of liver cancer has been reported for workers exposed to aflatoxins in concerns processing livestock feed [41]. Larsson and coworkers [42] have also shown that asymptomatic dairy farmers exposed to airborne mold dust may have signs of immunostimulation and inflammation in their alveolar space. Farmers exposed to mold dust may exhibit signs of alveolitis [42], and severe toxic irritative reactions can occur after a single inhalation of high levels of spores [43]. Studies have suggested that inhalation exposure to mold spores is another cause of organic dust toxic syndrome [44].

Occupational biohazards of biological origin are grouped into (1) occupational diseases of the respiratory tract and skin caused by allergenic/and or toxic agents forming bioaerosols, and (2) agents causing zoonoses and other infectious diseases spread through various exposure vectors [45].

3.2.1. *Allergenic and/or toxic agents*

A wide range of impacts may lead to different types of allergies. Substances such as microbial enzymes for food processing (e.g., α-amylase in commercial bakeries) and detergent are potent allergens that can cause asthma and rhinitis [24]. Many fungal species detected in bioaerosols in the food industry, for example, from the genera *Penicillium, Aspergillus,* and *Cladosporium* [46, 47], are responsible for respiratory disease and allergies in other environments [48]. Fungi produce copious amounts of spores that are easily dispersed in polluted air and dust [21]. The genera *Alternaria, Cladosporium, Aspergillus, Penicillium,* and *Fusarium* are more prone to cause sensitivity. Fungal allergy often appears as type I immediate, IgE-mediated hypersensitivity. In the case of allergic reaction, it can manifest as rhinitis or conjunctivitis, asthma, urticaria, or atopic dermatitis. This is called a type II hypersensitivity reaction as is the case in response to the mannan–polysaccharide of the cell wall of *Candida* and *Aspergillus*. An example of type III hypersensitivity is allergic alveolitis and bronchopulmonary aspergillosis [21]. Allergy to *Aspergillus fumigatus* is common in atopic asthma. In a large part of the population, allergies occur in the form of rhinitis, also accompanied by ocular signs [21]. It is estimated that approximately 2–6% of the general population in developed countries is allergic to fungi.

3.2.2. *Infection*

Recently, infectious diseases are being considered the most frequently occurring occupational diseases. Occupational biohazards are infectious agents or hazardous biological materials that exert harmful effects on workers' health, either directly through infection or indirectly through damage to the working environment, and it can also include medical waste or samples of a microbe, virus, or toxin from a biological source [45]. Most of the agents responsible for respiratory infections are spread through the air, primarily from person to person (anthroponoses), from living (zoonoses), the abiotic environment (e.g., soil and water), and decaying plant or animal matter (sapronoses) [24]. Inhalation is the most important and efficient route by which infectious agents enter the human body, and infections contracted by this route are the most difficult to control. Transmission by air allows an infectious agent to reach a larger number of potential hosts than would be possible if infected individuals had to come into direct contact to transfer microbes from person to person [24].

4. Legislation

Insufficient occupational exposure limits (OELs) set by regulatory organizations and the diversity of agents in occupational environments often complicate proper risk assessment of exposure to bioaerosols. Regulatory OELs have been adopted for cotton, grain, wood, flour, organic dust, and subtilisins (**Table 2**) [49, 50]. However, these limits are based on dust levels only and do not take specific components present in the dust into consideration. With the exception of subtilisin, even the OEL for "particulates not otherwise regulated" serves as reference where OELs are not specified [49]. Furthermore, the scientific evidence for certain set of exposure limits, such as ≈100 cells m^{-3} allowed for fungal and actinomycetes, can be difficult to access [51, 52]. In some cases, the risk of infectious agents and guidance on health surveillance and containment levels are provided [53], but no limits are specified for either infectious or noninfectious biological agents.

Specific OELs are required to protect workers' health. However, bioaerosol research has thus far only resulted in proposed exposure limits for endotoxins and fungal spores. A criteria document based on inflammatory respiratory effects [51] proposed a lowest observed effect level (LOEL) of 10^4 spores m^{-3} for nonpathogenic and nonmycotoxin-producing fungal species. Several organizations have also proposed guidelines for fungi in indoor environments, but the criteria were developed for assessing indoor mold problems and are not health based [54, 55]. For other agents, risk assessment may be based on exposure–response associations

Agent	ACGIH	Norway
Raw cotton dust	0.2 mg m^{-3}	0.2 mg m^{-3}
Grain dust (oat, wheat, barley)	4 mg m^{-3}	None
Flour dust	0.5 mg m^{-3}	3 mg m^{-3}
Wood dust	0.5 mg m^{-3}	1–2 mg m^{-3}
Organic dust	None	5 mg m^{-3}
Particulates not otherwise regulated	10 mg m^{-3}	10 mg m^{-3}

Table 2. Regulatory occupational exposure limits (OELs) for cotton, grain, wood, flour, organic dust and subtilisin.

found in relevant epidemiological studies, e.g., β (1→3)-glucans and allergens, but lack of standardization of measurement methods represents a great challenge [56, 57].

There are no uniform international standards available on levels and acceptable maximum bioaerosol loads (**Table 3**) [22]. The American Conference of Governmental Industrial Hygienists (ACGIH) stated that "a general threshold limit value (TLV) for culturable or countable bioaerosol concentrations is not scientifically supported" based on the lack of data describing exposure-response relationships [71]. New revised ACGIH will be released early 2017. Furthermore, no uniform standardized method is available for the collection and the analysis of bacterial and fungal bioaerosols, which makes the establishment of exposure limits challenging. Still, neither air sampling techniques nor identification and cultivation methods have been internationally standardized, impeding, therefore, the prospect of data comparison.

Country	Number of culturable organisms as CFU m^{-3}			References
	Bacteria	Yeast	Total Bioaerosols	
Brazil		750		[58, 59]
Canada		150		[60]
China	2500–7000 (location dependent)			[61]
Finland	4500			[62]
Germany	10,000	10,000		[63, 64]
Korea			800	[65]
Portugal		500		[66]
Netherlands	10,000		10,000	[67]
Russia		2000–10,000 (species dependent)		[68]
Switzerland	10,000 (aerobic mesophilic) 1000 (Gram-negative)			[69, 70]
USA		1000		[71, 72]
European Union	10,000 (private home) 2000 (nonindustrial indoor location)	10,000 (private home) 2000 (nonindustrial indoor location)		[73]

Table 3. Acceptable maximum bioaerosol loads allowed for indoor air quality in different countries.

5. Bioaerosol detection: approaches and complications

Bioaerosol monitoring is a rapidly emerging area of industrial hygiene [74]. Measurements include especially microbes in both indoor (e.g., industrial, office, or residential) and outdoor (e.g., agricultural and general air quality) environments [7]. It is necessary to evaluate their presence quantitatively (by a count or a determination) and/or qualitatively (by identifying the genus and species) [31]. Each bioaerosol sample is unique, as its composition varies in time and space (abundance and diversity of species, quantity of inflammatory components such as endotoxins and β-D-glucans). This often leads not only to high variation between samples from the same workplace, which can be due to external factors, but also to the dynamic evolution of the colonized substrate and fast multiplication rate of microbes [11].

5.1. Available sampling methods

A wide variety of bioaerosol sampling equipment are available, and no standardized protocols have yet been established. There are two primary methods for microbial air sampling, namely passive and active monitoring. Passive monitoring, also referred to as settle plates or petri plates, requires petri dishes containing agar or Petrifilm™ that are opened and exposed to the air for specified periods of time. Microbes that settle out of the ambient air can then be determined qualitatively. The passive approach offers lengthy sampling periods at low cost but does not take into account air movement or airborne populations per volume of air and may miss critical microbes [75]. Active monitoring requires a microbial air sampler to force air onto or into collection media at a specific rate over a specified time period. This approach is less time consuming and better for areas with low microbial loads and allows for both quantitative and qualitative analyses. However, vigorous air movement may cause injury to vegetative cells [76]. Three approaches can be used for active monitoring: impaction, impingement, and filtration.

Impaction involves the use of an air pump to capture air over the surface of a petri dishcontaining agar. The airflow over the agar is controlled by slits or holes that are arranged to distribute the airflow evenly over the agar surface. Sampling equipment is easy to use, and the consumable costs are relatively low. Different sampler options are summarized in **Table 4**. Drawbacks may include loss of microbial cells viability due to impact stress and loss of recovery efficiency due to the failure of microbes to adhere to agar surfaces. Competition for growth and the influence of selective media choices should also be considered when planning a monitoring strategy [92]. Impaction is often the preferred active monitoring approach for bioaerosol sampling in the food processing environment.

Impingement of microbes in a liquid matrix requires particulate laden air to accelerate as it is drawn through the cassettes tapered inlet slit and directed toward a small slide containing the collection media, where the particles become impacted, and the airflow continues out the exit orifice. With this approach, it is possible to measure both the culturable and the nonculturable components of bioaerosols and is ideally suitable for aeromicrobiology studies because the liquid matrix can be divided for various analyses. Sampler options are listed in **Table 5**. Collection vials are often constructed from glass and can be easily damaged or broken. This approach tends to be expensive and may also present low capture rates, loss of collection fluid to evaporation and violent bubbling, low capture rate of virus-sized particles, and loss of cell viability [101].

Sampler	Information	Difficulty to use	Flow rate	References
Single-Stage Viable Andersen Cascade Impactor	• N6 microbial impactor • Meet the specifications of latest ACGIH Bioaerosol Committee • EPA, OSHA and FDA referenced • Sharp cutoff diameter of 0.65 μm	Easy to use	28.3 L. min^{-1}	[31, 59, 77–81]
Two-Stage Viable Andersen Cascade Impactor	• Multi orifice cascade impactor • Whenever size distribution is not required • When only respirable segregation or total counts are needed • 95–100% of viable particles above 0.8 μm	Easy to use	28.3 L min^{-1}	[82]
Six-Stage Viable Andersen Cascade Impactor	• Multi-orifice cascade impactor • Measure the concentration and particle size distribution of aerobic bacteria and fungiViable particles can be collected on a variety of bacteriological agar • Calibrated to collect all particles (physical size, shape or density) • Can be directly related to human lung deposition	Easy to use	28.3 L min^{-1}	[31, 61, 83–90]
Mattson Garvin Slit-to-agar	• Accurate and quantitative • Sampling even the smallest of viable particles • Collection on 150 mm × 15 mm disposable culture plate • No dilution or plating steps are required • Results are expressed as viable particles per unit of air • Time-concentration relationship may be determined	Self-contained	cu ft min^{-1}	[31]
SAS Super 180	• Considered the international standard for portable air microbiology sampling • Pharmaceutical, food industry, hospital sector and indoor air quality • Used onboard the International Space Station	Easy to use	60–100 L min^{-1}	[86, 91, 92]
Biotest RCS	• Evaluate microbiological quality of ambient air, functionality of air treatment equipment and systems, effectiveness of decontamination measures • Collection on agar media strip	Pushbutton operation Remote control	50 L min^{-1}	[31]

Sampler	Information	Difficulty to use	Flow rate	References
IOM Sampler	• Reusable two-part filter cassette with specified 25-mm filters • Collection of inhalable airborne particles • Available in conductive plastic or stainless steel • Stainless steel model ideal for sampling vapor-phase isocyanates followed by chemical analysis • Sample culturable and nonculturable • Collection on membrane filters	Difficult to use	2 L min^{-1}	[37, 93]
SKC BioStage®	• Single stage • Viable cascade impactor • Meets NIOSH requirements and ACGIH recommendations • Collection on standard-size agar plates • SureLock positive seal ensures sample integrity	Easy to use	28.3 L min^{-1}	[31, 90, 92, 94]
SAMPL'AIR™	• 99% microbial collection rate • High efficiency, even with the small-est particles • Ideal for regular, thorough air quality control	Easy to use	100 L min^{-1}	[92]
MAS-100eco	• Sieve impaction systems • Accurately regulates airflow in real time • Collection media: 90–100 mm petri dish or 55–60 mm contact plate	Flexibility Remote control	100 L min^{-1}	[95–99]
RCS	• Rotary centrifugal air sampler • Lightweight and portable • Collection on agar strips	Easy to use	40 L min^{-1}	[34]

Table 4. Available impaction-based bioaerosol sampling devices.

Filtration involves pumping air through a porous membrane filter to capture bioaerosols. This method can be used to detect both culturable and nonculturable components and has proven highly efficient in trapping of microbes larger than the chosen pore size of the filter surface. It does, however, require expensive sampling equipment and sample processing, and data analysis may require a high level of expertise [102]. Available cassettes for filtration sampling of bioaerosols are listed in **Table 6**.

Sampler	Information	Difficulty to use	Flow rate	References
All-Glass (AGI-30) Impinger	• High velocity impinger • Can be used in heavily contaminated environments • Sampling times up to 30 min (dilute impinge solution prior to use)	Easy to use	12–13 L min^{-1}	[31]
Burkard May-Impringer	• Since 1966 • Fractions collected gently into liquid where clumps separate into viable units • Little danger of sample overload • Subsamples permit the use of a variety of culture methods • Particle fractions (>10 μm, 10–4 μm, <4 μm)	Difficult to use	20 L min^{-1}	[31]
BioSampler®	• Collection time up to 8 h with sonic-flow Vac-U-Go Sampler • Recommended for: infection control investigation in hospitals and veterinary clinics, biological research, infectious disease investigations in public buildings, and safety concerns in the food handling industry	Easy to use	12.5 L min^{-1}	[31]
Air-O-Cell® cassette	• Use with any standard off-the-shelf area sampling pump (15 LPM open flow) • Unique design for the rapid collection of a wide range of airborne aerosols including mold spores, pollen, insect parts, skin cell fragments, fibers (e.g., asbestos, fiberglass, cellulose, clothing fibers, etc.) and inorganic particulate, e.g., ceramic, fly ash, copy toner and so on). • Collects both viable and nonviable sample specimens • Direct microscopic analysis can be performed immediately • Collection media compatible with a wide range of biological stains and refractive index oils • Direct quantitative analysis of organic and inorganic particulate possible • Suitable for use in confined or restrictive spaces	Easy to use	15 L min^{-1}	[100]
Micro-Orifice Uniform Deposition Impactors™ (MOUDI™)	• 18 μm cut-point inlet stage • Additional stages to size-fractionate aerosols particles: 8-stage (0.18 μm) and 10-stage (0.056 μm)	Difficult to use	30 L min^{-1}	[101]

Table 5. Available impingement-based bioaerosol sampling devices.

Sampler	Information	Difficulty to use	Flow rate	References
Burkard Spore Trap (1,7-Day)	• Particles sizes 1–10 μm • Continuous sampling • Spores are impacted on adhesive coated transparent plastic tape (Melinex) • Sensitive to small changes in wind direction	Reliable and simple operation	10 L min^{-1}	[31]
Button Aerosol Sampler	• Porous curved-surface inlet • Particles sizes 100 μm	Easy to use	4 L min^{-1}	[31, 37]
Buck BioAire™ Model B520	• Compact, lightweight, controlled flow sampling pump • Uses Allergenco-D™ or Air-O-Cell™ cassettes • Unattended timed programming • 5 h of continuous operation	Easy to use	15 L min^{-1}	[103]
Zefon 37 mm Clear Styrene Air Sampling cassettes	• Meet all applicable NIOSH, OSHA and EPA air sampling standards	Easy to use	4 L min^{-1}	[37]
NIOSH Personal Bioaerosol Cyclone Sampler	• Tube wall impaction • Third stage filtering	Convenient Easy to use	4 L min^{-1}	[37, 93]

Table 6. Available filtration-based bioaerosol sampling devices.

5.2. Complications and considerations related to bioaerosol detection

It is important to emphasize that bioaerosols are ubiquitous environmental contaminants and in the majority of cases, not an integral part of the process. It would therefore be inappropriate to "sample-to-see-what-is-in-the-air" since the presence of microbes in the air can be expected. The field is dominated by lack of consistent data and an abundance of speculation [7]. The lack of standard methods, environmental guidelines, and databases complicates the interpretation and comparison of results [92]. Also, since no single method can fully characterize all bioaerosols components [7], it is imperative to do a proper evaluation/investigation before choosing a sampling method or initiating a sampling protocol. The following questions summarize important aspects to address when planning a bioaerosol monitoring approach and can be used as guidelines.

Why sample? Formulate the objectives for sampling clearly. It is important to establish whether sampling bioaerosols is necessitated by baseline monitoring for compliance or to confront an existing quality (product) and/or safety (food handler health) problem for which bioaerosols as causative agent need to be ruled out.

Where to sample? The notion of sampling before doing a critical assessment of the facility is a current shortcoming. This approach can even be misleading because it produces information that is difficult to interpret, might create unnecessary concern, and may lead almost inevitably to the sampling having to be repeated professionally/by external consultants. Foci for the assessment should include environmental factors, factory design/layout, equipment, product type, and food handlers (health, shifts/placement, skills level, training, behavior) [76]. Certain environmental factors such as temperature, airflow, and relative humidity can be associated with bioaerosol levels [104]. Heating, air-conditioning, or ventilating systems may provoke fluctuations in temperature and relative humidity. Detectable bacterial and fungal levels can also be affected by these factors, since they require specific environmental conditions to grow and propagate. Sampling sites to consider include areas with negative air pressure, raw material area where a lot of dust is generated, under air vents, areas where water spraying or misting can occur, active floor drains and areas with higher worker activity or other movement.

Which bioaerosol component to measure? Information from the evaluation/investigation should be able to establish which bioaerosol component is of interest: viable microbial components (culture dependent) or nonviable but still bioactive (culture independent) component. Although culture-dependent methods are by far the most widely used procedures for assessing the microbiological content of bioaerosols (**Table 1**); it is now widely accepted that such methods significantly underestimate the total quantity of microbes present. Plate count media describe the well-known problem that only a small fraction (10%) of airborne microbes forms colonies on a typical culture media, thus leading to a significant underestimation of the actual viable airborne bioaerosol concentration. The vast remaining number of airborne microbes can be described as viable but nonculturable, indicating very low metabolic activity or resting dormant state. Dead airborne bacteria or fungi debris or toxins retain their allergenic or toxic properties and are therefore also relevant to any occupational health assessment.

Which air sampler to use? Impingement sampling devices (**Table 5**) can be used to detect both viable and nonviable bioaerosol components. Either viable or nonviable components can be assessed using impaction (**Table 4**) or filtration (**Table 6**), respectively. Choosing a sampling device will also depend on availability, level of expertise and funding.

How often and when to sample? In a new program for compliance monitoring, it is advisable to start with more frequent data collection as this will allow for baseline establishment. When the data are available to show that the bioaerosols in a system/area are stable enough, the number of data collection points can be reduced. Microbial results can differ depending on the activity in a specific area. Sampling times should include both "dynamic" and "static" conditions monitoring.

Author details

Shirleen M. Theisinger and Olga de Smidt*

*Address all correspondence to: odesmidt@cut.ac.za

Centre for Applied Food Security and Biotechnology (CAFSaB), Central University of Technology, Free State, South Africa

References

[1] Jaenicke R. Abundance of cellular material and proteins in the atmosphere. Science. 2005;**308**:73. DOI: 10.1126/science.1106335

[2] Morris CE, Sands DC, Bardin M, Jaenicke R, Vogal B, Leyronas C. Microbiology and atmospheric processes: An upcoming era of research on bio-meteorology. Biogeosciences Discussions. 2008;**5**:191-212. DOI: 10.5194/bgd-5-191-2008

[3] Basinas I, Sigsgaard T, Erlandsen M, et al. Exposure affecting factors of dairy farmers' exposure to inhalable dust and endotoxin. Annals of Occupational Hygiene. 2014;**58**: 707-723. DOI: 10.1093/annhyg/meu024

[4] Pasteur L. Annales des Siences Naturelles. Zoologie et Biologie Animale. 1861;**44**:5-98

[5] Pepys J, Jenkins PA, Festenstein GN, Gregory PH, Lacey ME, Skinner FA. Lancet. 1963;**2**: 607-611

[6] Rylander R. Chest. Bacterial Toxins and the Ethology of Byssinosis. 1987;**79**:34S-38S

[7] Yoo K, Lee TK, Choi EJ, Yang J, Shukla SK, Hwang S, Park J. Molecular approaches for detection and monitoring of microbial communities in bioaerosols: A review. Journal of Environmental Sciences. 2016;**51**:234-247. DOI: 10.1016/j.jes.2016.07.002

[8] Sutton GHC. Enumeration of total airborne bacteria, yeast and mold contaminants and identification of *Escherichia coli* O157:H7, *Listeria* Spp., *Salmonella* Spp., and *Staphylococcus* Spp. in a beef and pork slaughter facility [thesis]. University of Florida, Gainesville; 2004

[9] Hirst JM. Handbook of Bioaerosols: Introduction, Retrospect and Prospect. Lewis Publishers, United States of America; 1995. pp. 1-10. ISBN: 0873716159, 9780873716154

[10] Shale K, Lues JFR. The etiology of bioaerosols in food environments. Food Reviews International. 2007;**23**:73-90. DOI: 10.1080/87559120600998205

[11] Oppliger A. Advancing the science of bioaerosol exposure assessment. Annals of Occupational Hygiene. 2014;**58**(6):661-663. DOI: 10.1093/annhyg/meu042

[12] Shaffer BT, Lighthart B. Survey of culturable airborne bacteria at four diverse locations in Oregon: Urban, rural, forest, and coastal. Microbial Ecology. 1997;**34**:167-177. DOI: 10.1007/s002489900046

[13] Brandl H, Fricker-Feet C, Ziegler D, Mamdal J, Stephan R, Lehner A. Distribution and identification of culturable airborne microorganisms in a Swiss milk processing facility. Journal of Dairy Science. 2014;**97**:240-246. DOI: 10.3168/jds.2013-7028

[14] Reddy MK, Srinivas T, Lakshmi A. A study of bioaerosols in indoor air of food godowns of Visakhapatnam, India. Journal of Environmental Research and Development. 2012;**6**: 447-451

[15] Tsai M, Liu H. Exposure to culturable airborne bioaerosols during noodle manufacturing in Central Taiwan. Science of the Total Environment. 2009;**5**:1536-1546. DOI: 10.1016/j.scitotenv.2008.10.029

[16] Vela J, Hildebrandt K, Metcalfe A, Rempel H, Bittman S, Topp E, Diarra M. Characterization of *Staphylococcus xylosus* isolated from broiler chicken barn bioaerosol. Poultry Science. 2012;**91**:3003-3012. DOI: 10.3382/ps.2012-02302

[17] Aringoli EE, Cambiagno DE, Chiericatti CA, Basilico JC, Basilico MLZ. Mycoflora study in a wheat flour mill of Argentina. Brazilian Journal of Microbiology. 2012;**43**:1444-1451. ISSN: 1517-8382

[18] Desai MR, Ghosh SK. Occupational exposure to airborne fungi among rice mill workers with special reference to aflatoxin producing A. Flavus strains. The Annals of Agricultural and Environmental Medicine. 2003;**10**:159-162

[19] Nakhchian H, Yazdi FT, Mortazavi SA, Mohebbi M. Isolation, identification and growth's comparison of mold types in a cake factory environment and final products. International Journal of Advanced Biological and Biomedical Research. 2014;**2**:2505-2517. ISSN: 2322-4827

[20] Agranovski E. Aerosols—Science and Technology. Wiley-VCH Verlag GmbH & Co. KGaA Haboken, New Jersey, USA; 2011. p. 69. ISBN: 978-3-527-32660-0

[21] Zukiewicz-Sobczak WA. The role of fungi in allergic diseases. Postepy Dermatologii I Alergologii. 2013;**1**:42-45. DOI: 10.5114/pdia.2013.33377

[22] Mandal J, Brandl H. Bioaerosols in indoor environment—a review with special reference to residential and occupational locations. The Open Environmental & Biological Monitoring Journal. 2011;**4**:83-96. DOI: 10.2174/1875040001104010083

[23] Kobayashi T, Iijima K, Redhakrisjnan S, et al. Asthma-related environmental fungus, Alternaria, activates dendritic cells and produce potent Th2 adjuvant activity. Journal of Immunology. 2009;**182**:2502-2510. DOI: 10.4049/jimmunol.0802773

[24] Ruzer LS, Harlet NH. Aerosol Handbook — Measurement, Dosimetry and Health Effects. 2nd ed. Boca Raton, FL, USA: CRC Press; 2012

[25] PirieI RS, Collie DD, Dixon PM, et al. Inhaled endotoxin and organic dust particulates have synergistic proinflammatory effects in equine heaves (organic dust-induced asthma). Clinical Immunology. 2003;**115**:797-802. DOI: 10.1046/j.1365-2222.2003.01640

[26] Madsen AM. Airborne endotoxin in different background environments and seasons. The Annals of Agricultural and Environmental Medicine. 2006;**13**:81-86

[27] Douwes J, Thorne P, Pearce N, et al. Bioaerosol health effects and exposure assessment: Progress and prospects. Annals of Occupational Hygiene. 2003;**47**:187-200

[28] Spaan S, Heederik DJJ, Thorne PS, Wouters IM. Optimization of Airborne Endotoxin Exposure Assessment: Effects of Filter Type, Transport Conditions, Extraction Solutions, and Storage of Samples and Extracts. Applied and Environmental Microbiology. 2007;**73**:6134-6143 [PubMed: 17675430]

[29] Sorensen WG, Jones W, Simpson J, Davidson JI. Aflatoxin in respirable airborne peanut dust. The Journal of Toxicology and Environmental Health. 1984;**14**:525-533. DOI: 10.1080/15287398409530603

[30] Straumfors HA, Nordy KC, Sletner KS, Elen O, Clasen PE, Eduard W. Toxigenic *Fusarium* spp. as determinant of trichothecene mycotoxins in settled grain dust. Journal of Occupational and Environmental Hygiene. 2010;**44**:1980-1986. DOI: 10.1080/15459620600987431

[31] Goyer N, Lavoie J, Lazure L, Marchand G. Bioaerosols in the Workplace: Evaluation, Control and Prevention Guide. Montreal: IRSST; 2001

[32] Pearson C, Littlewood E, Douglas P, Robertson S, Timothy WG, Hansell AL. Exposure and health outcomes in relation to bioaerosol emissions from composting facilities: A systematic review of occupational and community studies. The Journal of Toxicology and Environmental Health. 2015;**18**:43-69. DOI: 10.1080/10937404.2015.1009961

[33] Adhikari A, Gupta J, Wilkins JR, Olds RL, Indugula R, Cho KJ, Li C, Yermakov M. Airborne microorganisms, endotoxin, and (1-3)-β-D-glucan exposure in greenhouses and assessment of respiratory symptoms among workers. The Annals of Occupational Hygiene. 2010;**55**:272-285. DOI: 10.1093/annhyg/meq082

[34] Hargreaves M, Prappukkaran S, Morawska L, Hitchins J, Congrong H, Gilbert D. A pilot investigation into associations between indoor airborne fungal and non-biological particle concentrations in residential houses in Brisbane. Science of the Total Environment. 2003;**312**:89-101. DOI: 18.8ed96b68.1485663101.5a1752b

[35] Stetzwnbach LD, Buttner MP, Cruz P. Detection and enumeration of airborne contaminants. Current Opinion in Biotechnology. 2004;**15**:170-174. DOI: 18.8ed96b68.1485663390.5a2185e

[36] Nerin C, Azar M, Carrizo D. Food contamination during food process. The Journal of Food Science and Technology. 2016;**48**:63-68

[37] Wang C, Chen BT, Han B, Liu AC, Hung P, Chen C, Chao HJ. Field evaluation of personal sampling methods for multiple bioaerosols. PloS ONE. 2015;**10**(3):e120308, DOI: 10.1371/journal.pone.0120308

[38] Hansen VM, Meyling NV, Winding A, Eilenberg J, Madsen AM. Factors affecting vegetable growers' exposure to fungal bioaerosols and airborne dust. The Annals of Occupational Hygiene. 2012;**56**:170-182. DOI: 10.1093/annhyg/mer090

[39] Dvorackova I. Aflatoxin inhalation and alveolar cell carcinoma. British Medical Journal. 1976;**1**:691

[40] Hayes RB, van Nieuwenhuize JP, Raatgever JW, ten Kate FJ. Aflatoxin exposures in the industrial setting: An epidemiological study of mortality. Food and Chemical Toxicology. 1984;**22**(1):39-43. DOI: 18.8ed96b68.1485665440.5a68517

[41] Olsen JH, Dragsted L, Autrup H. Cancer risk and occupational exposure to aflatoxins in Denmark. British Journal of Cancer. 1988;**58**:392-396

[42] Larsson K, Malmberg P, Eklund A, Belin L, Blaschke E. Exposure to microorganisms, airway inflammatory changes and immune reactions in asymptomatic dairy farmers.

Bronchoalveolar lavage evidence of macrophage activation and permeability changes in the airways. International Archives of Allergy and Immunology. 1988;**87**(2):127-133. DOI: 10.1159/000234662

[43] Poulsen OM, Breum NO, Ebbehoj N, Hansen AM, Ivens UI, van Lelieveld D, et al. Sorting and recycling of domestic waste. Review of occupational health problems and their possible causes. Science of the Total Environment. 1995;**168**(1):33-56. DOI: 18.add9 6b68.1485665559.7f55015e

[44] Vogelzang PF, van der Gulden JW, Folgering H, van Schayck CP. Organic dust toxic syndrome in swine confinement farming. American Journal of Industrial Medicine. 1999;**35**:332-334. DOI: 10.1002/(SICI)1097-0274(199904)35

[45] Rim KT, Lim CH. Biologically hazardous agents at work and efforts to protect workers health: A review of recent reports. Safety and Health at Work. 2014;**5**:43-52. DOI: 18.add 96b68.1485665803.7f59fa23

[46] Chang CW, Chung H, Huang CF, Su HJJ. Exposure of workers to airborne microorganisms in open-air swine houses. Applied and Environmental Microbiology. 2001;**67**:155-161. DOI: 10.1128/AEM.67.1.155-161.2001

[47] Lee SA, Adhikari A, Grinshpun SA, McKay R, Shukla R, Reponen T. Personal exposure to airborne dust and microorganisms in agricultural environments. The Journal of Occupational and Environmental Hygiene. 2006;**3**:118-130. DOI: 10.1080/15459620500524607

[48] Flannigan B, Samson RA, Miller JD. Microorganisms in Home and Indoor Work Environments. London and New York: Taylor & Francis; 2001

[49] Teixeira JP, Gaspar J, Coelho P, Costa C, Pinho-Silva S, Costa S, Da Silva S, Laffon B, Pasaro E, Rueff J, et al. Cytogenetic and DNA damage on workers exposed to styrene. Mutagenesis. 2010;**25**(6):617-621. DOI: https//doi.org/10.1093/mutage/geq,049

[50] Norwegian Labour Inspection Authority. Guidance for administrative standards for contamination of the work environment. Regulations of 22 December 2014 No. 1883. Trondheim, Norway: The Labour Inspection Authority; 2011

[51] Eduard W. Real-time PCR detection of toxigenic *Fusarium* in airborne and settled grain dust and associations with trichothecene mycotoxins. Arbete och Halsa. 2006;**21**:1-145. Available from: http://gupea.ub.guse/dspace/bitstream/2077/4359/1/ah2006_21.pdf

[52] Sidorov KK. Maximum Allowable Concentrations of Harmful Substances in Workplace Air. Toksikologiceskij Vestnik, July 1, 1993. State committee for Hygiene and Epidemiological Surveillance. 1993:38-44

[53] Directive 2000/54/EC of the European parliament and of the council on the protection of workers from risks related to exposure to biological agents at work. Official Journal of the European Communities. 2000. Available from: http://eurlex.europa.eu/LexUriServ/LexUriServ.do?uri=OJ:L:2000:262:0021:0045:EN:PDF

[54] Rao CY, Burge HA, Chang JC. Review of quantitative standards and guidelines for fungi in indoor air. Journal of the Air & Waste Management Association. 1996;**46**:899-908

[55] WHO. Guidelines for Indoor Air Quality: Dampness and Mould. WHO Regional Office for Europe. 2009. Available from: www.euro.who.int/_data/assets/pdf_file/0017/43325/E92645.pdf

[56] Douwes J. (1→3)-β-D-glucans and respiratory health: A review of the scientific evidence. Indoor Air. 2005;**15**:160-169. DOI: 10.1111/j.1600-0668.2005.00333

[57] Heederik D. Health based occupational exposure limits for high molecular weight sensitizers: how long is the road we must travel. The Annals of Occupational Hygiene. 2002;**46**:439-446

[58] de Aquino Neto FR, de Goes Siqueira LF. Guidelines for indoor air quality in offices in Brazil. Proceedings of Healthy Buildings. 2004;**4**:549-554

[59] Nunes ZG, Martins AS, Altoe ALF, et al. Indoor air microbial evaluation of offices, hospitals, industries and shopping centers. The Memórias do Instituto Oswaldo Cruz. 2005;**100**:351-357. DOI: 10.1590/S0074-02762005000400003

[60] Bratlett KH, Lee KS, Stephens G, Black W, Brauer M, Copes R. Evaluating indoor air quality: Test standards for bioaerosol. Reports to the Workers' Compensation Board of British Columbia Research Secretariat. 2003;**99**:FS-64

[61] Gorny RL. Bilogiczne czynniki szkodliwe: Normy, zalecenia I propozycje wartosci dopuszczalnych. Podstawy I Metody Oceny SRodowiska Pracy. 2004;**3**:17-39

[62] Nevalainen A. Bacterial aerosols in indoor air [dissertation]. Helsinki, Finland: National Public Health Institute; 1989

[63] Institut fur Arbeitsschutz de Deutschen Gesetzlichen Unfallversicherung (IFA). Verfahren zur Bestimmung der Schimmelpilzkonzentration in der Luft am Arbeitsplatz (9420#). Berlin, Germany: Erich Schmidt Verlag; 2001

[64] Institut fur Arbeitsschutz de Deutschen Gesetzlichen Unfallversicherung (IFA). Verfahren zur Bestimmung der Bakterienkonzentration in der Luft am Arbeitsplatz (9430#). Berlin, Germany: Erich Schmidt Verlag; 2004

[65] Jo WK, Seo YJ. Indoor and outdoor bioaerosol levels at recreation facilities, elementary schools, and homes. Chemosphere. 2005;**61**:1570-1579. DOI: 18.add96b68.1485670753.7fdeefa8

[66] Pegas PN, Evtyugina MG, Alves CA, et al. Outdoor/indoor air quality on primary schools in Lisbon: A preliminary study. Quimica Nova. 2010;**33**:1145-1149. DOI: 10.1590/S0100-40422010000500027

[67] Heida H, Bartma F, van der Zee SC. Occupational exposure and indoor quality monitoring in a composting facility. American Industrial Hygiene Association Journal. 1995;**56**:39-43. DOI: 10.1080/15428119591017295

[68] Eduard W. Fungal spores: A critical review of the toxicological and epidemiological evidence as a basis for occupational exposure limit setting. Critical Reviews in Toxicology. 2009;**39**:799-864. DOI: 10.3109/10408440903307333

[69] SUVA. Grenzwerte am Arbeitsplatz. Luzern, Switzerland; 2007

[70] Oppliger A, Rusca S, Charriere N, Duc, TV, Droz PO. Assessment of bioaerosols and inhalable dust exposure in Swiss sawmills. The Annals of Occupational Hygiene. 2005;**49**:385-391. DOI: 10.1093/annhyg/meh105

[71] American Conference of Governmental Industrial Hygienists (ACGIH). Threshold limit values (TLVs) for chemical substances and physical agents and biological exposure indices (BEIs). Cinnati, USA; 2009. pp. 223-226

[72] Occupational Safety and Health Administration (OSHA). Indoor Air Quality Investigation. Technical Manual (OTM), Section II, Chapter 2; 2008

[73] Wanner HU, Verhoff A, Colombi A, et al. Biological particles in indoor environments. European Collaborative Action "Indoor air quality and its impact on man". Commission of the European Communities, Report No. 12. Luxembourg; 1994

[74] Jensen PA, Schafer MP. Sampling and characterization of bioaerosols. NIOSH manual of analytical methods. 1998.

[75] Moberg L, Kornacji JL. Microbiological monitoring of the food processing environment. In: Compendium of Methods for the Microbiological Examination of Foods. 5th ed, Chapter 3. Washington, DC: American Public Health Association; 2015

[76] Kornacki J. An environmental sample approach to product risk assessment. Food Safety Magazine. February 2014

[77] Burge HA. Bioaerosol investigations. In: Burge HA, editor. Bioaerosols. Boca Raton, USA: CRC Press; 1995. pp. 1-23

[78] Chao HJ, Schwartz J, Milton DK, Burge HA. Populations and determinants of airborne fungi in large office buildings. Environmental Health Perspectives. 2002;**110**:777-782

[79] Li CS, Hou PA. Bioaerosol characteristics in hospital clean rooms. Science of the Total Environment. 2003;**305**:169-176. DOI: 18.8ed96b68.1485673905.5c82266

[80] Kim KY, Kin CN, Kim D. Distribution characteristics of airborne bacteria and fungi in the general hospitals of Korea. Industrial Health. 2010;**48**:236-243. DOI: 10.2486/indhealth.48.362

[81] Scott JA, Summerbell RC, Green BJ. Detection of indoor fungi bioaerosols, in fundamentals of mold growth in indoor environments and strategies for healthy living. In: Fundamentals of mold growth in indoor environments and strategies for healthy living. Adan OCG, Samson RA, editors. Wageningen, The Netherlands: Wageningen Academic Publishers; 2011

[82] Zhu H, Phelan PE, Duan T, et al. Experimental study of indoor and outdoor airborne bacterial concentrations in Tempe, Arizona, USA. Aerobiologia. 2003;**19**:201-211. DOI: 10.1023/B:AERO.0000006571.23160.8a

[83] Dillon Hk, Heinsohn, PA, Miller JD. Field Guide for the Determination of Biological Contaminants on Environmental samples. Fairfax, VA: AIHA Publications; 1996

[84] Gorny RL, Dutkiewicz J. Bacterial and fungal aerosols in indoor environment in Central and Eastern Countries. The Annals of Agricultural and Environmental Medicine. 2002;**9**:17-23

[85] Gilbert Y, Veillette M, Duchaine C. Airborne bacteria and antibiotic resistance genes in hospital rooms. Aerobiologia. 2010;**26**:185-194. DOI: 10.1007/s10453-010-9155-1

[86] Benasconi C, Rodolfi M, Picco AM, Grisoli P, Dacarro C, Rembges D. Pyrogenic activity of air to characterize bioaerosols exposure in public buildings: A pilot study. Letters in Applied Microbiology. 2010;**50**:571-577. DOI: 10.1111/j.1472-765X.2010.02831.x

[87] Nasir ZA, Colbeck I. Assessment of bacterial and fungal aerosol in different residential settings. Water, Air, and Soil Pollution. 2010;**211**:367-377. DOI: 10.1007/s11270-009-0306-3

[88] Niesler A, Gorny RL, Wlazlo A, et al. Microbial contamination of storerooms at the Auschqitz-Birkenau Museum. Aerobiologia. 2010;**26**:125-133. DOI: 10.1007/s10453-009-9149-z

[89] Wang CC, Fang GC, Kuo CH. Bioaerosols as contributors to poor air quality in Taichung City, Taiwan. Environmental Monitoring and Assessment. 2010;**166**:1-9. DOI: 10.1007/s10661-009-0980-z

[90] Zhenqiang X, Kai W, Yan W, Fangxia S, Qi C, Mingzhen L, Maosheng Y. Enhancing bioaerosol sampling by Andersen impactors using mineral-oil spread agar plate. Plos ONE. 2013. DOI: 10.1371/journal.pone.0056896

[91] Osimani A, Aquilanti L, Tavoletti S, Clementi F. Microbiological monitoring of air quality in a university canteen: An 11-year report. Environmental Monitoring and Assessment. 2013;**185**:4765-4774. DOI: 10.1007/s10661-012-2903-7

[92] Therkorn JH, Mainelis G. Effect of agar plate volume on accuracy of culturable bioaerosol impactors. Aerosol Science and Technology. 2013;**47**:1353-1362. DOI: 10.1080/02786826.2013.842954

[93] Lecours PB, Veillette M, Marsolais D, Duchaine C. Characterization of bioaerosols from dairy barns: Reconstructing the puzzle of occupational respiratory diseases by using molecular approaches. Applied and Environmental Microbiology. 2013;**78**:3242-3248. DOI: 10.1128/AEM.07661-11

[94] Yao M, Mainelis G. Analysis of potable impactor performance for enumeration of viable bioaerosols. Journal of Occupational and Environmental Hygiene. 2007a;**4**:514-524. DOI: 10.1080/15459620701407388

[95] Kalogeraski N, Paschali D, Lekaditis V, Pantidou A, Eleftheriadis K, Lazaridis M. Indoor air quality—Bioaerosol measurements in domestic and office premises. Journal of Aerosol Science. 2005;**36**:751-761. DOI: 18.add96b68.1485678314.80eb53dd

[96] Haas D, Habib J, Galler H, et al. Assessment of indoor air in Austrian apartments with and without visible mold growth. Atmospheric Environment. 2007;**41**:5192-5201. DOI: 18.8ed96b68.1485678843.6003d66

[97] Brandl H, von Daniken A, Hitz C, Krebs W. Short term dynamic patterns of bioaerosols generation and distribution in an indoor environment. Aerobiologia. 2008;**14**:203-209. DOI: 10.1007/s10453-008-9099-x

[98] Zorman T, Jersek B. Assessment of bioaerosol concentrations in different indoor environments. Indoor and Built Environment. 2008;**17**:155-163

[99] Uztan AH, Ates M, Abaci O, et al. Determination of potential allergenic fungal flora and its clinical reflection in suburban elementary schools in Izmir. Environmental Monitoring and Assessment. 2010;**168**:691-702. DOI: 10.1007/s10661-009-1144-x

[100] Godwin C, Batterman S. Indoor air quality in Michigan schools. Indoor Air. 2007;**17**:109-121. DOI: 10.1111/j.1600-0668.2006.00459.x

[101] Urbano R. Palenik B, Gaston CJ, Prather KA. Detection and phylogenetic analysis of coastal bioaerosols using culture dependent and independent techniques. Biogeosciences. 2011;**8**:301-309

[102] Chen YP, Cui Y, Dong JG. Variation of airborne bacteria and fungi at Emperor Qin's Terra-Cotta Museum, Xi'an, China, during the "Oct 1" Gold week period of 2006. Environmental Science and Pollution Research International. 2010;**17**:478-485. DOI: 10.1007/s11356-009-0161-1

[103] Rittenour WR, Ciaccio CE, Barnes CS, Kashon ML, Lemons AR, Beezhold DH, Green BJ. Internal transcribed spacer r RNA gene sequencing analysis of fungal diversity in Kansas City indoor environments. Environmental Science: Processes and Impacts. 2014;**16**:33-43. DOI: 10.1039/C3EM00441D

[104] Srikanth P, Sudharsanam S, Steinberg R. Bio-aerosols in indoor environment: Composition, health effects and analysis. The Indian Journal of Medical Microbiology. 2008; **26**:302-312

Chapter 4

Some Aspects of Animal Feed Sampling and Analysis

Gabriel Adebayo Malomo and
Nnemeka Edith Ihegwuagu

Additional information is available at the end of the chapter

http://dx.doi.org/10.5772/intechopen.70856

Abstract

Animal feed plays an important part in the food chain and the composition and quality of the livestock products (milk, meat and eggs) that people consume. Animal feeds are either classified as fodder, forage, or mixed feeds. Fodders could be classified as roughages (fresh cut forage, hay or dry forage, straw, root crops, stover and silage) and concentrates such as grains, legumes and by-products of processing. Safety is perhaps one of the most important reasons for feed analysis by the manufacturers and consumers. Storage duration and conditions for feed samples, as well as of stable and unstable parameters are important in sample preparation. A number of sub-samples for preparing final sample for various categories of feed products are recommended. Some analysis conducted on feed include; dry matter, crude ash, ash insoluble in acid (sand), crude protein, crude fat, fibre analysis, starch, gross energy, minerals. More are amino acids (excluding tryptophan), amino acids (tryptophan), fatty acids, vitamins, reducing sugar, mycotoxins, and pesticides. Various types of samples depending on their purposes and uses are available from check, standard, working and referee samples to composite types. Sampling errors in procedures exists and can be minimized by standards or purposes of the analysis, appropriate sampling equipment and using the right quantity of materials.

Keywords: animal, feed, sampling, analysis, quality control

1. Introduction

Food is any substance, originating from plants, animal or any other source, consumed by any organism for the purpose of providing nutritional support. When consumed and assimilated, food is used in the body to maintain and repair body tissues, promote health and growth, sustain life, provide energy, for reproduction and other vital body processes through the release of its nutrients. Essentially, the basic nutritive components of food are carbohydrates, proteins, fats, minerals, vitamins and water, which are absorbed in the body in various usable forms.

Food given to food-producing animals, whether made up of single or multiple materials, are generally referred to as feed or feedstuff, and could be fed as raw, semi-processed or processed [1]. Feeds may be live organisms, particularly in the production of aquatic organisms. Animal feeds are either classified as fodder, forage, or mixed feeds. Fodders could be classified as roughages (fresh cut forage, hay or dry forage, straw, root crops, stover and silage) and concentrates such as grains, legumes and by-products of processing. Plant materials consumed by grazing animals either directly as pasture, crop residue, and immature cereal crops are referred to as forage. However, forage materials cut as fodder, particularly fresh, hay, and silage are sometimes loosely referred to as forage. Mixed feeds are produced from several feed ingredients combined in different proportions to achieve a particular nutritional quality. Feed ingredients, including additives, may or may not add any nutritional value to the mixed feed and comprises of components originating from plant, animal, or aquatic sources, which could be organic or inorganic in nature [1]. Several ingredients used for the production of feedstuff are limiting in one or more nutrients, and must therefore be blended in appropriate proportions to meet the nutritional requirements of the animals. Mixed feeds are usually produced in the form of mash or pellets.

Animal feeds are important, not only to the feeds manufacturers and animal producers, but also to the regulators, policy makers, processors and the final consumers of the end-products. This is because animal feed is an integral part of the food supply chain and it is critical to the efficient and profitable production of quality and safe food. Thus, feed safety is critical to food safety. Stakeholders interested in producing safe foods must be, and are rightly, concerned with the safety of animal feeds. Research evidence regarding risks associated with consumption of contaminated feeds and several epidemics which were traceable to animal feeds in different countries have made the demands for safe feed even more serious in recent times [2]. Ingredients, suppliers and processing methods used in the process of feed production may significantly impact public health [3]. As part of the measures to ensure that feed ingredients and feedstuffs meet the various quality and safety requirements, a wide range of analyses, both scientific and socio-economic, are carried out in the feed and food industry. Some of the reasons stakeholders carry out feed analysis are for regulations and enforcement, recommendations, labelling, validation of manufacturers' quality claims, feed/food safety and defence, quality control in feed production, and for research and development. Adoption of standard sampling and analytical methods assist in accurately characterizing the problems and contribute to the integrity of the results. This chapter focuses on some aspects of analysis in the feed industry to ensure the production of nutritious and safe food animals.

2. Animal feed and food supply chain

Every step from primary production to final consumption, that is, from farm to fork, makes up the food chain. Feed production, plays significant role in the production of food of animal origin and it is, therefore, a critical aspect of the food chain (**Figure 1**). Therefore, all key actors on every nodes of the food chain are responsible for the production of safe, healthy and nutritious feeds.

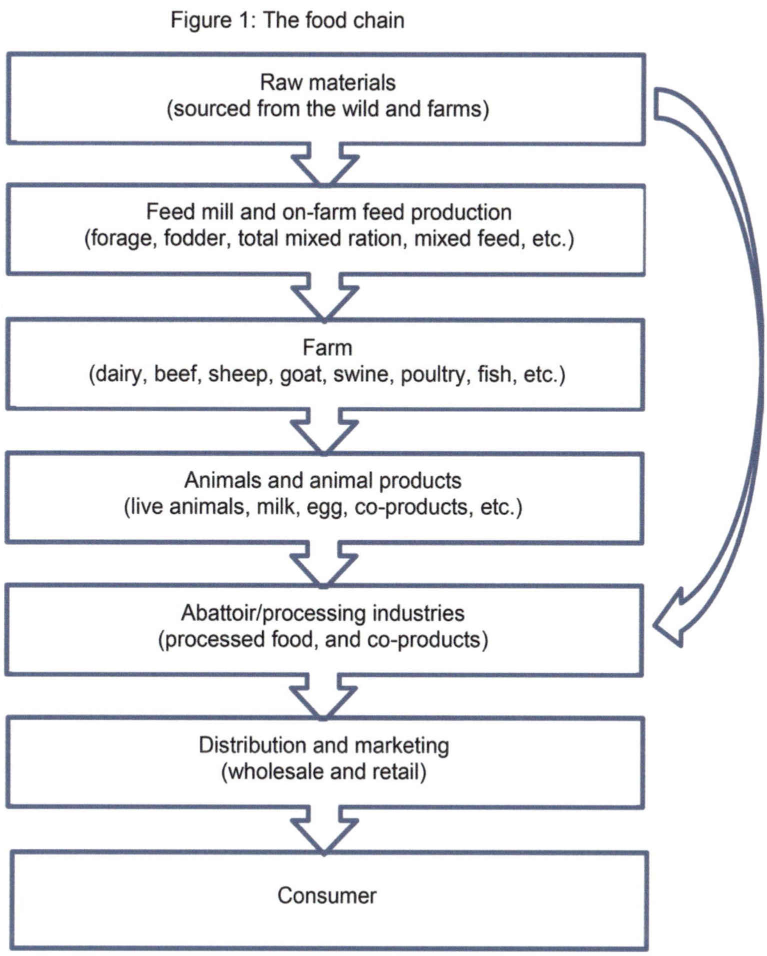

Figure 1. The food chain.

3. Feed hazards

Food safety hazards connected with feed safety has heightened the concern and level of seriousness given globally to feed analysis in recent times. Contaminants in feeds can be inherent or naturally occurring such as mycotoxins and heavy metals, or industrial substances such as polychlorinated biphenyls (PCBs) and pesticides [4]. Feed contaminants, as in the case of food contaminants, can also be biological, chemical, and physical (radionuclides) [5, 6]. Three important criteria used for selecting hazards of current importance in feeds are relevance of hazards to public health; extent of occurrence of the hazard; and impact of the hazard on international trade in food and feed [7].

Codex Alimentarius standards for contaminants in feeds and foods applies to substances with food and feed significance but no public health significance, pesticides residues, residues of veterinary drugs and feed additives, microbial toxins, and residuals of processing aids [8]. Related with each hazard are specific sources and routes of contamination and exposure, which may be deliberately or accidentally introduced along the feed production value chain. Examples of sources of hazards that may be present in animal feeds and feed ingredients are presented in **Table 1**.

It is worthy of note that among a wide range of sources, feedstuffs especially those of plant origin are the most common potential means of exposing animals to toxic levels of minerals [23]. However, feedstuff of animal origin can contain potentially toxic levels of some minerals (**Table 2**). Toxic levels of minerals in feedstuff, which may lead to death of the animals, may occur as a result of utilizing feed ingredients sourced from areas with high concentrations of heavy metals, processing methods, feed formulation and manufacturing errors, and contamination during storage or transportation.

Type of hazard	Causative substance/agents	Sources in animal feeds	Analytical method
Physical hazards	Glass, metals, plastic and wood	Handling at various stages of production and processing	Physical inspection
Chemical Hazards	Dioxins, dibenzofurans, dioxin-like PCBs	Contaminated mineral sources, food by-products, fish by-products	Gas chromatography-high resolution mass spectrometry (GC/HR-MS); gas chromatography with other lowre solution mass spectrometry instruments; Calux-assay methods
	Mycotoxins – Aflatoxin B1, ochratoxin A, zearalenone, fumonisin B1, deoxinivalerol, T-2, HT-2	Cereals (especially maize), cotton seed, peanut (groundnut), copra, distillers' dried grains with soluble (DDGS)	Semi-quantitative ISO method based on thin-layer chromatography [11] and a methods applying HPLC with fluorimetric detection after immuno-affinity clean-up; dipstick-like immunochemical screening methods are also applied; Official methods of [12].
	Veterinary drugs	Terrestrial and aquatic-based feed ingredients, medicated feeds, DDGS	HPLC methods; Enzyme-Linked ImmunoSorbent Assay (ELISA); Microbiological inhibition assays; LC-MS/MS or liquid chromatography with diode array detector (LC-DAD) methods; official AOAC method [13].
	Organopesticides – DDT, hexachlorobenzene and aldrin.	Contaminated feed ingredients and feeds	GC-MS; GC with electron-capture detection (ECD) methods
Microbial contaminants	Brucella, salmonella, endoparasites (Echinococcus, *Toxoplasma gondii*, Cisticercus and Trichinella)	Contaminated pasture, forages, and animal and vegetable protein meals.	Methods of [11, 14–22]

Source: Adapted from Refs. [9, 10].

Table 1. Sources and analytical methods for detection of some chemical and microbiological hazards in feeds.

Mineral	Major sources	Animal health concerns	Analytical method	Difficulties with analysis
Undesirable heavy metals				
Arsenic	Sea plants, fish products; and supplemental minerals.	Medium	Hydride Atomic absorption spectrophotometry (AAS); Plasma mass spectrometry; Graphite Furnace AAS; and Silver diethyldithiocarbamate colorimetric methods.	Incomplete extraction; Retention time irreproducibility; Co-elusion of species; Presence of unidentified species; Lack of standards; and Detection interference.
Cadmium	Mineral supplements such as phosphate, zinc sources; Forage/grains (depending on geographical area); Manure, sewage sludge, or phosphate fertilizer enriched soil or biosolids.	High	AAS; or inductively coupled atomic emission spectroscopy methods. Neutron activation analysis or X-ray fluorescence in living animals.	Susceptibility to contamination and Detection interference
Lead	Contaminated soil, lead paints, water from plumbing systems that contain lead, batteries. Mineral supplements (copper sulphate, zinc sulphate, zinc oxide).	High	Flame atomic absorption spectrometry (AAS), graphite furnace atomic absorption spectrometry (GFAAS), anode stripping voltametry (ASV), inductively coupled plasma-atomic emission spectroscopy (ICP/AES), inductively coupled plasma mass spectrometry, and X-ray fluorescence spectroscopy methods.	Lead contamination during sample collects
Mercury/ methyl mercury	Anthropogenic contamination, fish meal.	High	Cold vapour AAS, atomic fluorescence spectrometry (AFS), electrothermal atomic absorption (ETAAS), neutron activation analysis (NAA), mass spectrometry (MS), and anodic stripping voltammetry (ASV) methods. Capillary gas-liquid chromatography with electron-capture method is used to determine methyl mercury levels in biological samples	Quantification of each species is important; Relative volatility and loss during sample storage, preparation and analysis; oxidizing properties of lab ware can lead to loss of methyl mercury; contamination with mercury; repeated freezing and thawing of wet biological samples can lead to loss of methyl mercury
Other heavy metals				
Copper	Gras and leguminous forages; cereal grains, leguminous oil seed meals; Poultry and swine wastes; Mineral supplements (cupric sulphate, tribasic cupric chloride, copper oxide (primarily cupric oxide), cupric carbonate, and various organic copper sources).	High	Atomic absorption spectrophotometric (AAS) methods (flame or graphite furnace). Inductively coupled plasma-atomic emission spectroscopy (ICP-AES).	Contamination during sample collection

Mineral	Major sources	Animal health concerns	Analytical method	Difficulties with analysis
Iron	Alfalfa; Cereal grains. Leguminous and oil seeds; meat meals; fish meals; Mineral supplements (Ground limestone, oyster shell, and many forms of calcium Phosphate); iron sulphate, iron chloride, iron proteinates, and blood meal.	Medium	Flame AAS, graphite furnace AAS, ICP-AES, ICP-MS, and X-ray fluorescence spectroscopy methods.	Iron contamination during sample collection and processing; Contamination from the atmosphere during analysis if sample is not covered and analysed in a hood.
Zinc	Pasture herbage; cereal grains; leguminous meals; fish meal; whale meal; meat meal; mineral supplements	Medium	Atomic absorption spectrophotometry and inductively coupled plasma emission spectrophotometry	Volatilization loss during ashing at >500°C; contamination of samples
Chromium	Feed grade monocalcium phosphate and defluorinated phosphate Sources	Low	Graphite furnace atomic absorption spectrometry, neutron activation analysis, or mass spectrometry methods are used for low concentrations. Flame atomic absorption spectrometry and inductively coupled plasma-atomic emission spectrometry are used for potentially toxic levels in feedstuff	Background and environmental contamination of biological samples with chromium during collection, storage, and preparation of samples for analysis can present a major source of error. Losses through volatilization during heating or acid digestion of samples.
Molybdenum	Marine origin soil; alkaline soils; pasture; Sodium molybdate	High	Colorimetric; atomic absorption spectroscopy (Graphite furnace AAS); inductively coupled plasma-atomic emission spectrometry methods	Interferences from ferric iron and tungsten' artificially elevated level due to molybdenum carbides accumulation on the wall of graphite furnace
Selenium	Plants from selenium-rich soil	High	Fluorometry; Atomic absorption spectrometry; improved atomic absorption method based on Zeeman effect background; Neutron activation analysis (NAA); HPLC-ICP-MS (inductively coupled plasma mass spectrometry) or HPLC-ESI (electronspray ionization)-MS Methods	Volatility and instability of certain forms of selenium and the non-homogeneity of sample materials; contamination with selenium during sample collection, preparation and storage; explosion hazard; extended digestion time may be required for some samples like urine, some plants and kidney tissue; equipment scarcity for NAA method;

Source: Adapted from NRC [23].

Table 2. Heavy metals, sources and analytical methods.

In view of the potentially harmful effects of toxic levels of elemental minerals in animal feeds, it is essential for stakeholders to have a good grasp of the maximum tolerable levels (MTL) so that it will not be exceeded. Maximum tolerable limits of a mineral is the dietary level that will not impair the performance or health of the animal when fed over a particular period of time [23]. In 2005, National Research Council's Committee on Minerals and Toxic Substances in Diets and Water for Animals recommended MTL of some minerals in feeds of food-producing animals and may serve as a guide depending on the condition of the animals.

4. Methods of analysis in the feed industry

Types of analyses conducted by laboratory are proximate analyses, macro-minerals, micro-minerals at trace level, chromatographic analyses (such as amino acids, fatty acids, etc.) and chromatographic analyses at trace level (contaminants such as aflatoxins, pesticides and pesticide residues, antibiotics, etc.) [24]. Several standard and laboratory methods have been developed over the years for the detection of both nutrients and contaminants in feed ingredients and feedstuffs. Garfield [25] classified the methods into official methods (required by law and used by regulatory and complying organization), reference methods (developed by collaborating organizations for validation purposes), screening or rapid methods (usually for large samples to determine whether further analysis are required with more accurate methods), routine methods (can be official, standard or modified methods used for routine testing), automated methods (may be official or screening methods that adopts automated equipment), and modified methods (usually official or standard methods, which have been modified to make it simple and applicable to wide range of samples).

In the absence of standardized analytical methods, laboratory methods that meet certain criteria, validated and accredited in line with international guidelines and quality assurance protocols, may serve as alternatives. 'Accuracy, applicability (matrix and concentration range), limit of detection, limit of determination, precision, repeatability and reproducibility' are some of the criteria that laboratory methods must meet be serve as an alternative to standard methods [10]. Analytical methods for detection of chemicals, including micro-minerals at trace levels and contaminants in feed ingredients and feedstuff have been highlighted in **Tables 1** and **2**.

4.1. Proximate analyses

Characterization of feeds and feed ingredients for general nutritional parameters are done using proximate analyses. The ability to conduct proximate analyses is the minimum requirement for laboratories [24]. Proximate analyses can be conducted in any basic nutrition laboratory while other analyses can be done in more complex laboratories. Analytical methods for proximate composition and some other feed components are presented in **Table 3**.

4.2. Risk analysis

Demands for higher standards in all aspects of feed production have been on the increase globally. This may be in part due to the increasing awareness of the role of feeds in potential

Parameters	Description
Dry matter	Part of the sample that remains after dying at 103°C
Crude ash	Part of the sample that remains after incineration at 550°C
Ash insoluble in acid (sand)	Ash that remains after boiling in strong acid
Crude protein	Total nitrogen content and to calculate the protein content by multiplying the nitrogen content by an appropriate conversion factor (usually ×6.25). Kjeldahl method (Nitrogen is converted into ammonia which is absorbed in boric acid and titrated against a standard acid); Dumas method (With complete combustion of sample at 950°C in the presence of oxygen, nitrogen is converted to a gaseous state and reduced to N2, followed by measurement in a thermal conductivity cell)
Crude fat	Non-polar extractable fraction of the sample. The extraction can be performed with or without prior acid hydrolysis, both being complementary methods. The laboratory should offer both options
Fibre analysis	Digestion of feed directly in the detergent solution and filtration *using crucibles* (official standard method). Digestion of sample whilst in a *nylon bag* and then washing the bag containing the digested sample to make it detergent free.
Starch	Starch can be measured by the classical Ewers method or with an enzymatic method. The enzymatic method can be used for all sample types and is therefore preferable
Gross energy	Gross energy represents the total energy value of the sample and is measured by bomb calorimeter.
Minerals	Minerals are generally measured by spectrometric methods following incineration and hydrolysis.
Amino acids (excluding tryptophan)	The standard method for the determination of amino acids is based on the hydrolysis of protein to amino acids using a strong acid with or without previous oxidation, followed by chromatographic separation and detection after derivatization
Amino acids (tryptophan)	Determination of tryptophan is based on an alkaline hydrolysis followed by chromatographic Separation
Fatty acids	The standard method for fatty acids is based on isolation and derivatization, followed by gas chromatographic separation
Vitamins	Determination of individual vitamins is based on extraction, followed by clean-up, concentration if needed, and chromatographic measurement.
Reducing sugar	Reducing sugars contain the most important sugars, including glucose, fructose and sucrose. Determination is based on the Luff-Schoorl principle.
Mycotoxins	Mycotoxins are undesirable substances produced by fungi (moulds). These present a potential danger to animal and human health. The maximum levels are nationally and internationally regulated. The different methods are based on extraction, purification, chromatographic separation and detection.
Pesticides	Pesticides are undesirable substances whose maximum levels are defined in national and international regulations. These regulations demand a low detection limit and positive identification of the pesticides, which is achieved by using mass spectrometric detection. The methods are based on extraction, purification, derivatization, chromatographic separation and identification.

Source: de Jonge and Jackson [24].

Table 3. Description of typical tests in feed analyses.

hazards associated with food of animal origin. Accordingly, appropriate codes have been developed by relevant international bodies to assist national authorities to take measures that would mitigate most of these risks, particularly those of public health importance and which may constitute barriers to international trades. Risk analysis is an objective and defensible mechanisms for risks reduction that are associated with health and other factors. For example, Article 2.1 of the Aquatic Animal Health Code, which addresses animal health issues in international trades, provided basic guide and steps for import risk analysis in relation to aquatic animals and aquatic animal products [26]. However, the principles and methods of risk analysis are the same for both aquatic and terrestrial animals and products, including feedstuff. The four components involved with risk analysis are highlighted below:

a. Hazard identification: This is a categorisation step in the risk analysis and the risk assessment should be concluded at this stage in the absence of any identified potential risk.

b. Risk assessment: Involves both qualitative and quantitative methods of risk assessment, each with its relevant outputs. The steps are entry assessment; exposure assessment (both entry and exposure assessment steps involve the assessment of biological, country and commodity factors); consequence assessment (direct and indirect consequences); and risks estimation which integrates results of the entry, exposure and consequence assessments to produce the overall measures of risks associated with the hazard identified at the outset. The risk assessment should be concluded at either entry assessment or exposure assessment step if no substantial risk is demonstrated. The whole risks pathway from identified hazard to unwanted outcome is taken into account by the risk estimation step.

c. Risk management: This involves deciding and implementing protective measures and at the same time minimizing the negative effects on trade. Components of risk management include risk evaluation, option evaluation, implementation, and monitoring and review

d. Risk communication: This requires having a risk communication strategy in place at the outset of each risk analysis.

4.3. Quality assurance and control in feed analysis

Variations in the results of feed analyses obtained from different laboratories have been a major source of concern in the feed industry and among relevant authorities globally [27–30]. Efforts to limit unacceptably high variations in the results of analysed samples in various laboratories, which are sometimes difficult to attribute to genotypic, environmental or inter-laboratory differences, contributed to the development of quality assurance and control for analysis [31]. Use of quality assurance schemes, inter-laboratory evaluation programmes and reference materials were recommended by [32] to reduce errors due to laboratory and methodological differences. Laboratory quality assurance scheme requires the implementation of management quality policy statement, objectives of the scheme, control of samples and records, equipment maintenance, methods evaluation, measurement principles, training, methods selection, intra- and inter-laboratory testing, reference standards, field and lab sampling, statistical considerations, audits, corrective actions, programme revisions and update [7]. These could be grouped properly under the four guiding principles of valid analytical measurement (VAM), which was developed in 1994 in the United Kingdom by the Department of Trade and Industry to contribute to validity of analytical data, namely:

i. Use of properly validated methods of measurement.

ii. Incorporate certified reference materials (CRMs) in quality assurance protocols to ensure traceability measurements.

iii. Independent assessment of laboratory's performance for particular tests through participation in national and international proficiency testing schemes (PTS).

iv. Independent approval of quality assurance arrangements of laboratories by accreditation or licensing to a recognized quality standard.

5. Some aspects of and considerations in feed sampling

The accuracy and reliability of the results of any analysis in the feed industry begins with the quality of sampling. An analysis can be said to be as good as its sampling because several challenges that can affect accuracy and reliability of the results are associated with sampling of the feeds and feed materials [4]. It is, therefore, critical to ensure sampling of feed ingredients and feeds is done in an area and in a way that makes the procedures easy, minimize the risk of contamination and cross contamination, makes proper performance of the laboratory analysis possible, and ensures all safety and health precautions for the sampler and the environment [7].

5.1. Types of samples

Pierce [33] identified various types of samples depending on their purposes and uses as follows: check sample; composite sample; discrete sample; duplicate sample; official sample; purchasing sample; referee sample; reference sample; retained sample; standard sample, and working sample.

5.2. Sampling errors

Sampling errors may be due to the heterogeneity of the inspected characteristics, the random nature of sampling, and the known and acceptable characteristics of the sampling plan [34].

Some of the measures to be taken to minimize sampling errors in the feed industry include

i. Sampling procedures should be based on the objectives, standards, or purposes of the analysis. Simple random sampling, stratified random sampling, and systematic sampling are examples of common sampling schemes used in the feed industry [35].

ii. Use appropriate sampling equipment that will not introduce contamination. For example, do not use lead containing materials to collect samples meant for lead analysis. Examples of sampling equipment include grain probes (slotted grain probes, open-handled grain probes, open-handled spiral probe); pelican grain sampler; tapered bag triers; double tube bag triers; single-tube, open-ended bag triers; bomb or zone sampler [35].

iii. Collect representative samples. If the samples collected are not representative of the whole, the results of the analysis become skewed. To collect a representative sample, the sampling scheme must be followed, adequate quantity of sample must be collected, and sampling equipment and procedure must be appropriate, required inspection of sample, among other things.

iv. Use the right quantity of materials and avoid splashing of samples during collection and analysis. Several errors can be associated with the splitting of samples, if not done carefully.

v. Use standard reference materials.

vi. Repeat analysis.

vii. Validate laboratory methodologies and use standard methods.

viii. Use well trained and knowledgeable personnel.

ix. Observe sampling precautions required for the methods of analysis.

x. Use the appropriate sampling plans.

5.3. Sampling plans selection

Sampling plan is a planned procedure that enables the choice of separate samples from a lot, for the purpose of getting the needed information, such as a decision on compliance status of a lot. It is also a scheme that defines the number of items to collect and the number of non-conforming items required in a sample to evaluate the compliance status of a lot [34]. Thus, without an appropriate sampling plan, it may be practically impossible to accurately decide the compliance status of a particular lot of a product. Codex guideline for sampling [34] recommends seven important considerations in selecting appropriate sampling plans in compliance with relevant standards in the feed industry: (i) existence (or not) of international reference document on sampling of the products under consideration; (ii) nature of control (individual or whole lot), (iii) nature of the characteristic to control (qualitative or quantitative characteristics), (iv) choice of the quality level, limiting quality or acceptance quality level, in line with principles laid down in Codex Manual of procedures and the type of risk, (v) nature of the lot, that is bulk or pre-packed products, size, homogeneity and distribution concerning the characteristics of control, (vi) composition of sample, that is those composed of single or more than one sampling unit, (vii) choice of the type of sampling plan.

5.4. Preparation of samples

Codex code [34] also sets the guidelines for sample preparation. A primary sample is prepared by direct collection of items or incremental samples. During the first stage of the sampling process, primary samples are collected from lots of items or incremental samples for pre-packed or bulk feeds, respectively. In order to facilitate laboratory analysis, sufficient quantity of the primary samples of similar size should be collected. Necessary precautions must be taken to ensure sample integrity and avoid any form of contamination throughout the entire process of sampling and analysis.

Composite sample is prepared, whenever required by the sampling plan, by carefully mixing the primary samples. This involves primary samples collected from a lot of pre-packaged products or incremental samples from a bulk (not-pre-packed) lot. In composite sample preparation, combination of primary samples may lead to loss of information on sample-to-sample variation. The composite sample should, except when too large, constitute the final sample which is sent to the laboratory for analysis (**Table 4**).

Product	Quantity in tons	Number of sub-samples	Minimum quantity of collective sample	Minimum quantity of final sample
Products in receptacles such as bags, drums, big bags, etc.				
Feed materials	Up to 50 tons	2	2 kg	300 g
	Above 50 tons	1 per 25 tons	1 kg per sub-sample	300 g
Compound feeds	All quantities	1	300 g	300 g
Premixes	All quantities	1	100 g	100 g
Feed additives	Up to 1 ton	2	250 g	100 g
	1–50 tons	2	1 kg	100 g
	Above 50 tons	1 per 25 tons	500 g per sub-sample	100 g
Products in storage tanks and silos or shed in the event of an emergency or accident				
Feed materials	Up to 50 tons	2	2 kg	600 g
	50–500 tons	1 per 25 tons	1 kg per 25 tons	600 g
	Part of the batch in excess of 500 tons	1 per 50 tons	1 kg per sub-sample	600 g
Compound feeds, premixes and feed additives	Up to 50 tons	2	2 kg	200 g
	50–500 tons	1 per 25 tons	1 kg per 25 tons	200 g
	Part of the batch in excess of 500 tons	1 per 50 tons	1 kg per sub-sample	200 g
Feed products in bulk per axle or during bagging				
Feed materials	Up to 50 tons	2	2 kg	300 g
Compound feeds	Up to 50 tons	1	300 g	300 g
Premixes	Up to 50 tons	1	100 g	100 g
Feed additives	Up to 50 tons	2	100 g	100 g
Forage products				
Forage products	Up to 50 tons	5 minimum	500 g	250 g
	Above 50 tons	10 minimum	500 g	250 g
Forage products in bulk, transport per axle				
Solid	Up to 50 tons	2 minimum	500 g	500 g
Feed products delivered by vessels or through water ways				
All products	Up to 5000 tons: for each 500 tons	5 minimum	Minimum of 1 kg for each 500 tons	300 g
	5000–10,000 tons for each 1,000 tons	5 minimum	1 kg for each 1,000 tons	300 g
	More than 10,000 tons for each 5000 tons	5 minimum	1 kg for each 5,000 tons	300 g

Source: GMP + International [36].

Table 4. Recommended number of sub-sample for preparing final sample for various category of feed products.

Tests samples are prepared from each composite sample by using appropriate grinding and crushing, sample division and mixing procedures. Some analytes or constituents may be degraded during the process of sample preparation due to a number of factors (**Table 5**).

5.5. Storage of feed samples

There are instances feed samples meant for laboratory analysis requires storage over a specified period of time. The recommended storage duration and conditions for feed samples are presented in **Table 6**.

Origin	Stable parameters	Unstable parameters	Reason(s) for degradation/change
Nutrients	(Crude) protein, fat, ash, fibre	Moisture	Temperature (volatile)
	Starch, sugar, lactose	Ammonia	Temperature (volatile)
	Gas production and enzyme-soluble organic substance production in *in vitro* tests	Organic acids (e.g. lactic acid, acetic acid, butyric acid, fumaric acid, formic acid)	Temperature (volatile)
	Minerals (e.g. Ca, P, Mg, Na, K, Cl)	Unsaturated fatty acids	Air oxidation (can result in production of short-chain fatty acids)
Feed additives	Trace elements (e.g. Cu, Zn, Mn, Fe, Se, Co)	Vitamins (e.g. vitamin A, C, D, E)	Temperature, ultraviolet (UV) light, air oxidation (sensitive)
	Amino acids (e.g. lysine, methionine, tryptophan)	1,2-Propanediol, ethylene glycol	Temperature (volatile)
	Enzymes (e.g. phytases, non-starch polysaccharide enzymes)	Microorganisms like probiotics (e.g. *Saccharomyces cerevisiae, Enterococcus faecium)*	Temperature (freezing), pressure (sensitive to grinding); moisture/dryness (influences growth of microorganisms)
Undesirable substances	Heavy metals (e.g. As, Pb, Cd, Hg)	Mycotoxins (e.g. aflatoxin B_1, deoxynivalenol, fumonisins, ochratoxin A, T-2 toxin, HT-2 toxin, zearalenone, ergot alkaloids)	Mould growth and change of mycotoxins possible at room temperature; UV light (sensitive – aflatoxin B_1)
	Dioxins and polychlorinated biphenyls (PCBs) with similar effects to dioxins	Drugs, antibiotics, pesticides	Temperature (sensitive)
		Hydrocyanic acid	Temperature (volatile)
Banned substances	Proteins of animal origin	Banned drugs, banned antibiotics	Temperature (sensitive)
(Other) Microorganisms		Yeasts, bacteria, moulds	Temperature (sensitive), dryness, influx of oxygen (anaerobiosis)

Source: ISO [37].

Table 5. General classification of stable and unstable parameters in relation to sample preparation.

Product	Storage duration	Storage conditions
Compound feeds (including milk replacer)	3–6 months	Cool, dry and dark
Premixes / processing aids	1 year or longer if there is still product in storage	Cool, dry and dark
feed additives	6 months	Cool, dry and dark
feed materials (dry, artificially dried, naturally dried)	6–12 months depending on the moment of delivery	Sample pot, cool, dry and dark
Fresh feed materials	Max 1 month, storage life often only a few days and will be fed as soon as possible	In air-tight sample bag in freezer
Preserved feed materials (products which are acidified or which have been subjected to natural acidification for the purpose of extending the shelf life of these products)	As long as the product is provided as feed up to a maximum of 2 years.	Preserved product (for example wrapped grass hay bale or green maize silage) is therefore "packaged", that it is available during the storage period for analysis.
Liquid and wet feed materials which are sensitive to decay due to their high moisture content	3 months or as long as it may be assumed that the product will be provided as feed.	In air-tight deepfreeze sample pot
Liquid and wet feed materials which are not sensitive to decay	3 months or as long as it may be assumed that the product will be provided as feed.	Sample pot, cool, dry and dark

Source: GMP + International [36].

Table 6. Storage duration and conditions for feed samples.

6. Conclusion

Food is deemed to be unsafe if it has an adverse effect on human health or it would make the food derived from food-producing animals unsafe for human consumption. Animal feed plays a critical role in the production of safe and nutritious food. There are several considerations that enhance quality and effective decision making in the feed and food production chain(s). Feed sampling and analyses are essential parts of the processes to ensure that feedstuffs and the resultant food animals meet all necessary standards. The reliability and quality of the analysis depends on the accuracy of sampling. Therefore adequate care must be taken to ensure that the analytes are handled in a way that will prevent degradation and errors. Where a feed which has been identified as not satisfying the feed safety requirement is part of a batch, lot or consignment of feed of the same class or description, it shall be presumed that all of the feed in that batch, lot or consignment is so affected, unless following a detailed assessment there is no evidence that the rest of the batch, lot or consignment fails to satisfy the feed safety requirement. This is an important point if you get an adverse sample result when sampling.

Author details

Gabriel Adebayo Malomo[1]* and Nnemeka Edith Ihegwuagu[2]

*Address all correspondence to: digabby1@gmail.com

1 Livestock Research Division, Agricultural Research Council of Nigeria (ARCN), Abuja, Nigeria

2 Natural Resources Management Division, Agricultural Research Council of Nigeria (ARCN), Abuja, Nigeria

References

[1] FAO/WHO. Code of Practice on Good Animal Feeding (CAC/RCP 54-2004). Rome: FAO; Geneva: WHO; 2004; 10 p

[2] Alkhalaf NA, Osman AK, Salama KA. Monitoring of aflatoxin and heavy metals in some poultry feeds. African Journal of Food Science. 2010;**4**(4):192-199

[3] FAO. Animal feeding and food safety. In: Report of an FAO Expert Consultation Rome, 10-14 March 1997. FAO Food and Nutrition Paper No. 69. Rome: FAO; 1998; 45 p

[4] Barel S. Review of the Needs and Composition of the Feed Safety and Noxious Compounds Inspection System in the Israeli VSAH. Bet Dagan: Israeli Veterinary Services and Animal Health; 2013; 154 p

[5] Maule AG, Gannam AL, Davis JW. Chemical contaminants in fish feeds used in federal salmonid hatcheries in the USA. Chemosphere. 2007;**67**(7):1308-1315. DOI: 10.1.1.492.9499

[6] Ihegwuagu NE, Emeje MO. Food quality control: History, present and future. In: Valdez B, editor. Scientific, Health and Social Aspects of the Food Industry. Croatia: InTech; 2012. p. 421-438. DOI: 10.5772/ 33151

[7] FAO and IFIF. Good practices for the feed industry – Implementing the Codex Alimentarius code of practice on good animal feeding. FAO Animal Production and Health Manual No. 9. Rome: FAO; 2010; 106 p

[8] FAO/WHO. General standard for contaminants and toxins in food and feed. In: Codex Stan 193-1995. Adopted in 1995. Revised in 1997, 2006, 2008, 2009 Amended in 2010, 2012, 2013, 2014, 2015, 2016

[9] FAO/WHO. Animal feed impact on food safety. In: Report of FAO/WHO Expert Meeting, FAO Headquarters, Rome, October 8-12. Rome: FAO; Geneva: WHO; 2008; 74 p. Available from: http://www.fao.org/3/a-a1507e.pdf [Accessed: 2017-08-23]

[10] Wicker DL. Feeding techniques to reduce the nitrogen content of poultry waste. In: Proceedings of the Multi-State Poultry Meeting, Indianapolis, IN, 1993; 18 p. Cited in: Patterson, P. 'Lesson 11: Using Dietary and Management Strategies to Reduce the Nutrient Excretion of Poultry', Iowa State University, Ames, Iowa 50011-3080. Iowa: MidWest Plan Service; 2001, 32 p

[11] AOAC, editor. Official Methods of Analysis. 17th ed. Gaithersburg, MD, USA: AOAC International; 2000

[12] ISO. Microbiology of Food and Animal Feeding Stuffs – Horizontal Method for the Detection of *Escherichia coli* O157 ISO 16654:2001. Geneva: International Organization for Standardization; 2001

[13] AOAC. Official Methods of Analysis. 18th ed. Gaithersburgs, MD: Association of Official Analytical Chemists; 2006

[14] ISO. Microbiology of Food and Animal Feeding Stuffs – Horizontal Method for the Detection of Salmonella spp., ISO 6579:2002. Geneva: International Organization for Standardization; 2002

[15] ISO. Microbiology of Food and Animal Feeding Stuffs – Horizontal Method for the Detection and Enumeration of Presumptive *Escherichia coli* – Most Probable Number Technique, ISO 7251:2005. Geneva, International Organization for Standardization; 2005

[16] ISO. Microbiology of Food and Animal Feeding Stuffs – Horizontal Method for the Detection of *Escherichia coli* O157, ISO 16654:2001. Geneva: International Organization for Standardization; 2005

[17] ISO. Microbiology of Food and Animal Feeding Stuffs – Horizontal Method for the Detection and Enumeration of Coliforms – Most Probable Number Technique, ISO 4831:2006. Geneva: International Organization for Standardization; 2006

[18] OIE. Bovine brucellosis. In: Manual of Diagnostic Tests and Vaccines for Terrestrial Animals, Chapter 2.3.1. 5th ed. Paris: World Organization of Animal Health; 2004

[19] OIE. Cysticercosis. In: Manual of Diagnostic Tests and Vaccines for Terrestrial Animals, Chapter 2.10.1. 5th ed. Paris: World Organization of Animal Health; 2004

[20] OIE. Salmonellosis. In: Manual of Diagnostic Tests and Vaccines for Terrestrial Animals, Chapter 2.10.3. 5th ed. Paris: World Organization of Animal Health; 2004

[21] USDA. Chapter 4a. Diarrheagenic *Escherichia coli*. In: Feng P, Weagant SD, editors. Bacteriological Analytical Manual Online. Washington DC: United States Food and Drug Administration; 2002

[22] USDA. Salmonella. In: Andrews WH, Hammack TS, editors. Bacteriological Analytical Manual Online (updated June 2006). Washington, DC: United States Food and Drug Administration; 2003

[23] NRC. Mineral Tolerance of Animals. Washington DC: National Research Council; 2005

[24] de Jonge LH, Jackson FS. The feed analysis laboratory: Establishment and quality control. Setting up a feed analysis laboratory, and implementing a quality assurance system compliant with ISO/IEC 17025:2005. In: Makkar HPS, editor. Animal Production and Health Guidelines No. 15. Rome: FAO; 2005

[25] Garfield FM. Quality Assurance Principles for Analytical Laboratories, Arlington, VA, USA: AOAC; 1984, 212 p

[26] OIE. Aquatic Animal Health Code. 17th ed., Paris: World Organization of Animal Health; 2014. ISBN: 978-92-9044-942-3

[27] Beever DE, Cammell SB, Edmonds S. Inter-laboratory variation in foodstuff evaluation of two contrasting maize silage samples. Animal Feed Science. 1996;**62**:685-686

[28] Thiex N, Smallidge R, Beine R. Sources of error in vitamin A analysis. Journal of AOAC International. 1996;**79**:1269-1275

[29] Key PE, Patey AL, Rowling S, Wilbourn A, Worner FM. International proficiency testing of analytical laboratories for foods and feeds from 1990 to 1996: The experiences of the United Kingdom food analysis performance assessment scheme. Journal of AOAC International. 1997;**80**:895-899

[30] Givens DI, Owen E, Adesogan AT. Current procedures, future requirements and the need for standardization. In: Givens DI, Owen E, Axford RFE, Ohmed HM, editors. Forage Evaluation in Ruminant Nutrition. Wallingford, UK: CABI Publishing; 2000. p. 449-474. DOI: 10.1079/9780851993447.0449

[31] Mueller-Harvey I. Modern techniques in feed analysis. In: Assessing Quality and Safety of Animal Feed. Rome: FAO; 2004. pp. 1-34

[32] Petterson DS, Harris DJ, Rayner CJ, Blakeney AB, Choct M. Methods for the analysis of premium livestock grains. Australian Journal of Agricultural Research. 1999;**50**:775-787

[33] Pierce JG. Sampling. In: Feed Manufacturing Technology IV. Arlington, VA, USA: AFIA; 1994

[34] FAO/WHO. General guidelines on sampling (CAC/GL 50-2004). Rome: FAO; Geneva: WHO; 2004

[35] Herrman T. Sampling: Procedure for Feeds. MF-2036 – Feed Manufacturing. Manhattan, Kansas, USA: Kansas State University; July, 2001

[36] GMP + International. Minimum Requirements for Sampling: GMP+ BA13. The Netherlands: GMP+ Feed Safety Assurance Scheme; 2012 25 p

[37] ISO. Animal Feeding Stuffs – Guidelines for Sample Preparation. EN ISO 6498:2012. Geneva: International Standard Organization; 2012; 46 p

Section 2

Sample Preparation Applications in Food and Beverage Analysis

Chapter 5

Analysis of Pesticide Residues in Chili (*Capsicum annuum* L.) using Ultra Performance Liquid Chromatography with UV Detection

Dinesh C Bilehal, Mahadev B. Chetti, Deepa G. T and Mahadev C. Khethagoudar

Additional information is available at the end of the chapter

http://dx.doi.org/10.5772/intechopen.70061

Abstract

The aim of this study was to analyze the pesticide residues in chili samples, collected from farmer's field. Ultra performance liquid chromatography (UPLC) with BEH C_{18} column was used for this analysis work. A cheap and fast method for the simultaneous quantification of 12 residue of pesticides in chili has been developed. Samples were prepared according to Quick, Easy, Cheap, Effective, Rugged, Safe (QuEChERS) method and quantification was performed by using tunable ultra violet (TUV) detector. The method was applied for the analysis of the chili samples and results showed that most of the samples have detectable pesticide residues. The residues of acetamiprid and thiodicarb were detected only in three samples, whereas flubendiamide and mancozeb were detected in six samples and arbosulfan and Spinosad were detected in two and five samples, respectively. Out of the 30 chili samples, only 11 samples were found to be contaminated with pesticide residues with more than maximum residue limits (MRLs).

Keywords: pesticide residues, ultra performance liquid chromatography (UPLC), QuEChERS, chili

1. Introduction

Chili [*Capsicum annuum* L.] is one of the major spice crop in India. Indian chilies have gained global demand due to high color value and low pungency [1]. The total world production of red chili is estimated to be around 21 lakh tons, 45% of which is produced in India [2]. The world spice production statistics records a bulk of 86% by volume, making the country the largest producer of spices, in addition to it being the largest consumer and exporter of spices in the global context [3, 4]. Chili has high medicinal value due to the abundance of availability of carotenoids, capsaicinoids [5], oleoresins, and mineral content [6]. Most of the studies have demonstrated that consumption of chili rich diets, increases in energy expenditure and oxidation of fat, and also it helps in the curing of many diseases [7].

Intensive agriculture practice receives most of the pesticides during different stages of cultivation. Pesticides increase crop productivity, reduce cost of production, improve quality, and thus help to increase in the farmers' income. The role and contribution of pesticides will be much more in the coming years, especially in the developing country like India. The demand for food continues to grow steadily due to growth of population. Although modern polar pesticides like organophosphorus and carbamates that replaced classical organochlorine pesticides are less persistent. There are more than 800 pesticide molecules used to control pests and also weeds [8, 9]. It is not possible to control the residues of pesticides in food commodities; hence, these compounds will accumulate in the human body after consumption through diets [10]. Hence, to overcome the effects of pesticides on different groups, the uniform maximum residue limits (MRL's) was established as 0.01 mg/kg for any pesticides [10].

In order to determine such a low level of detection of various analytes in the sample, a sophisticated instrument like gas chromatography (GC) or liquid chromatography (LC) have to be used for accurate separation and determination. With the advancement in the detectors in gas chromatography techniques namely electron capture detector (ECD), thermal conductivity detector (TCD), nitrogen phosphorus detector (NPD), and mass spectrometry detector (MSD), hence it is widely used in all analysis. Recently, polar and thermolabile pesticide analysis, liquid chromatography is used as alternative technique, where as these pesticides are not determinable by gas chromatography [11, 12]. For the analysis of wide range of polar pesticide residues in food commodities high-performance liquid chromatography mass spectrometry (HPLC–MS/MS) has become the important technique by choice [13].

Most of the published methods either expensive or involves laborious procedure for cleanup step during the extraction procedure, hence there is a chance of losing some quantity of analyte molecule. Similarly, some problems arise in the solvent exchange step, before applying the extract to the LC column, makes preparation of sample procedure less effective. Many challenges exists both in use of sophisticated equipments and sample handling procedure during pesticide residue analysis. In order to avoid such a complication in sample preparation, it is necessary to adopt Quick, Easy, Cheap, Effective, Rugged, Safe (QuEChERS) method. The ultra performance liquid chromatography (UPLC) is having more advantages than routine high-performance liquid chromatography (HPLC) system in terms of lesser retention time, resolution, and more sensitivity [14]. The UPLC separation was faster (six times) than regular HPLC system with monolithic column [15, 16]. And also, it consumes 80% of less mobile phase than normal HPLC system. The aim of the present study is to analyse the 12 pesticide residues with UPLC system using QuEChERS extraction method and critically determine the replacement of HPLC method with new UPLC method.

2. Experimental

2.1. Chemicals and materials

The certified reference materials (CRM's) of acetamiprid (purity 99%), benomyl (99%), flubendiamide (98.5%), indoxacarb (98.5%), carbosulfan (99%), imidacloprid (98%), methomyl (99%), thiodicarb (96%), spinosad (99%), oxydemeton-methyl (99%), difenoconazole (98.5%), and mancozeb (98.5%) for this study were obtained from Dr. Ehrenstorfer GmbH, Augsburg, Germany. HPLC grade solvents (acetonitrile, methanol, acetic acid, and formic acid) were obtained from Merck India Ltd. (Mumbai, India). Mobile phase water was prepared using millipore water purification system. Anhydrous sodium acetate and magnesium sulfate were procured from Sigma-Aldrich (Germany). And primary secondary amine (40 μm, Bondesil PSA) was purchased from Agilent Technologies (Bangalore, India).

2.2. Selection of pesticides

As many as 12 pesticides (**Table 1**) were used in this study, which are liquid chromatography amenable. And these pesticides are monitered in chili for the export to European Union. The pesticides chosen were those most often sprayed in chili cultivation.

2.3. Collection and storage of chili samples

Thirty chili samples (**Tables 2** and **3**) were collected randomly from different farmers' field of Haveri district, Karnataka, India. Two kilograms of each sample was taken, sealed in polythene bags, and stored at −4°C in deep freezer for further processing.

Pesticides	Retention time (RT)	Correlation coefficient (R^2)	Limit of detection (LOD) (mg/kg)	Limit of quantification (LOQ) (mg/kg)
Acetamiprid	2.544	0.9969	0.0010	0.0030
Benomyl	3.420	0.9971	0.0005	0.0015
Flubendiamide	3.802	0.9988	0.0005	0.0015
Indoxacarb	4.502	1.0000	0.5000	0.1500
Carbosulfan	5.975	1.0000	0.0005	0.0015
Imidacloprid	6.200	0.9986	0.0005	0.0015
Methomyl	6.431	0.9999	0. 0005	0.0015
Thiodicarb	6.556	0.9998	0.0005	0.0015
Spinosad	8.738	0.9999	0.0005	0.0015
Oxydemeton-methyl	8.997	0.9970	0.0005	0.0015
Difenoconazole	10.013	1.0000	0.0005	0.0015
Mancozeb	10.561	0.9999	0.0005	0.0015

Table 1. Retention time (RT), correlation coefficient (R2), limit of detection (LOD), and limit of quantification (LOQ) of 12 reference standards.

Name of pesticides	No. of chili samples (Residues in ppm)															
	MRLs prescribed by EU in ppm	**1**	**2**	**3**	**4**	**5**	**6**	**7**	**8**	**9**	**10**	**11**	**12**	**13**	**14**	**15**
Acetamiprid	0.30	**0.03**	ND	ND	**0.03**	ND	ND	ND	ND	ND	**0.04**	ND	ND	ND	ND	ND
Benomyl	0.10	ND	ND	ND	ND	ND	ND	ND	ND	ND	ND	ND	ND	ND	ND	ND
Flubendiamide	020	**0.20**	ND	ND	0.28	ND	ND	ND	ND	**0.35**	ND	ND	ND	ND	ND	**0.20**
Indoxacarb	0.30	ND	ND	ND	ND	ND	ND	ND	ND	ND	ND	ND	ND	ND	ND	ND
Carbosulfan	0.05	ND	ND	ND	ND	ND	ND	ND	ND	ND	ND	ND	ND	ND	**0.06**	ND
Imidacloprid	1.00	ND	ND	ND	ND	ND	ND	ND	ND	ND	ND	ND	ND	ND	ND	ND
Methomyl	0.02	ND	ND	ND	ND	ND	ND	ND	ND	ND	ND	ND	ND	ND	ND	ND
Thiodicarb	0.02	ND	ND	ND	**0.02**	ND	ND	ND	ND	ND	ND	ND	ND	ND	**0.02**	ND
Spinosad	2.00	**4.0**	ND	ND	**2.5**	ND	ND	ND	ND	**2.00**	ND	ND	ND	ND	ND	ND
Oxydemeton-methyl	0.01	ND	ND	ND	ND	ND	ND	ND	ND	ND	ND	ND	ND	ND	ND	ND
Difenoconazole	0.05	ND	ND	ND	ND	ND	ND	ND	ND	ND	ND	ND	ND	ND	ND	ND
Mancozeb	5.00	**5.0**	ND	ND	**5.1**	ND	ND	ND	ND	**5.0**	ND	ND	ND	ND	ND	**5.6**

Name of pesticides	No. of chili samples (Residues in ppm)															
	MRLs prescribed by EU in ppm	**16**	**17**	**18**	**19**	**20**	**21**	**22**	**23**	**24**	**25**	**26**	**27**	**28**	**29**	**30**
Acetamiprid	0.30	0.30	ND	ND	ND	ND	ND	ND	ND	ND	ND	ND	ND	ND	ND	ND
Benomyl	0.10	0.10	ND	ND	ND	ND	ND	ND	ND	ND	ND	ND	ND	ND	ND	ND
Flubendiamide	0.20	0.20	ND	ND	ND	ND	ND	ND	ND	**0.22**	ND	ND	ND	ND	ND	**0.20**
Indoxacarb	0.30	0.30	ND	ND	ND	ND	ND	ND	ND	ND	ND	ND	ND	ND	ND	ND

Name of pesticides	No. of chili samples (Residues in ppm)															
	MRLs prescribed by EU in ppm	**16**	**17**	**18**	**19**	**20**	**21**	**22**	**23**	**24**	**25**	**26**	**27**	**28**	**29**	**30**
Imidacloprid	1.00	1.00	ND	ND	ND	ND	ND	ND	ND	ND	ND	ND	ND	ND	ND	ND
Methomyl	0.02	0.02	ND	ND	ND	ND	ND	ND	ND	ND	ND	ND	ND	ND	ND	ND
Thiodicarb	0.02	0.02	ND	ND	ND	ND	ND	ND	ND	ND	ND	ND	**0.03**	ND	ND	ND
Spinosad	2.00	2.00	ND	ND	ND	**2.0**	ND	ND	ND	ND	ND	ND	ND	ND	ND	**2.2**
Oxydemeton-methyl	0.01	0.01	ND	ND	ND	ND	ND	ND	ND	ND	ND	ND	ND	ND	ND	ND
Difenoconazole	0.05	0.05	ND	ND	ND	ND	ND	ND	ND	ND	ND	ND	ND	ND	ND	ND
Mancozeb	5.00	0.05	ND	ND	ND	ND	ND	ND	ND	ND	ND	ND	ND	ND	ND	ND

ND = Not detected.

Table 2. Monitoring of pesticide residues in chili samples collected from farmers field of Haveri district, Karnataka using UPLC.

Sl. no	Name of pesticide	Number of positive samples	Incidence of residence (%)
1	Acetamiprid	3	10.00
2	Flubendiamide	6	20.00
3	Carbosulfan	2	6.66
4	Thiodicarb	3	10.00
5	Spinosad	5	16.66
6	Mancozeb	6	20.00

Table 3. Incidence of pesticide residues in 30 chili samples collected from farmer's field of Haveri district, Karnataka.

2.4. Preparation of reference standards

The individual stock solutions were prepared by exactly weighing 10 (±0.01) mg of certified reference standards in volumetric flask, dissolved in 10 ml methanol (1000 ppm), and were stored in a refrigerator −10 (±2)°C. Intermediate standards were prepared by diluting the stock solutions of 10 ppm and mix these with appropriate quantities for standard mixture preparation with acetonitrile. And these were stored at −10 (±2)°C and was used for 3 months. A working standard was prepared for diluting these intermediate stock solutions. Calibration plot was constructed using these standards.

2.5. Calibration

Five different standards of different concentrations like 500 ppt, 1 ppb, 10 ppb, 1 ppm, and 10 ppm were prepared using a serial dilution technique from 10 ppm concentration with acetonitrile as a solvent. For the same concentration levels, matrix matched standards were prepared in chili using the procedure mentioned in Section 2.6. Before doing this exercise, control chili samples were screened for the confirmation of absence of pesticide residues of the interest.

2.6. Sample preparation

Modified QuEChERS method was adopted for the preparation of the chili samples. The method involves crushing of 2 kg chili samples under ambient laboratory conditions. The 200 g of chili sample was further homogenized for 2 min and then 10 g of this sample were transferred in 50 ml polypropylene tubes and extracted with 10 ml acetonitrile (1% acetic acid)) in presence of 6 g anhydrous magnesium sulfate and 1.5 g sodium acetate. Then homogenization of the mixture was done at 15,000 rpm for about 2 min and centrifuged for 5 min at 6000 rpm. Dispersive solid phase extraction (d-SPE) was employed for the supernatant (1 ml) cleaning using 50 mg primary secondary amine (PSA) and 150 mg $MgSO_4$, which completely removes carbohydrates and fatty acids [17]. The supernatant was centrifuged at 3000 rpm for 5 min and the filtered through polyvinylidene difluoride (PVDF) membrane filter and transferred to auto sampler vial.

2.7. UPLC analysis

UPLC analysis was carried out using an ACQUITY UPLC™ system (Waters, USA), and separation was performed using Acquity UPLC BEH C18 (100 mm × 2.1 mm) with 1.7 μm particle size. The mobile phases used were (A) acetonitrile and (B) 0.1% formic acid. The gradient was linear from 0 to 30% A for 11 min and from 30 to 100% A for 1 min, followed by washing with B and re-equilibration of the column for 2 min were maintained for re-equilibration of the column to original state. The optimized parameters used were 0.2 mL/min flow rate, 45°C column temperature, and 25°C sample temperature and volume of injection was 1 μL throughout the analysis. Absorbances were recorded on-line at 280 nm using TUV detector.

3. Results and discussion

3.1. Optimization of chromatographic separation conditions

Mobile phase namely acetonitrile was used for the optimization of the system for the separation of reference standards using UPLC BEH C18 column. Generally, with change in the concentration of formic acid, the retention time of the individual standard varies. With the optimized gradient steps, we got good separation of the 12 standards with 0.1% formic acid (**Figure 1**). The optimum parameters used for this experiments were as follows: the mobile phase gradient was linear from 0 to 30% A for 11 min and from 30 to 100% A for

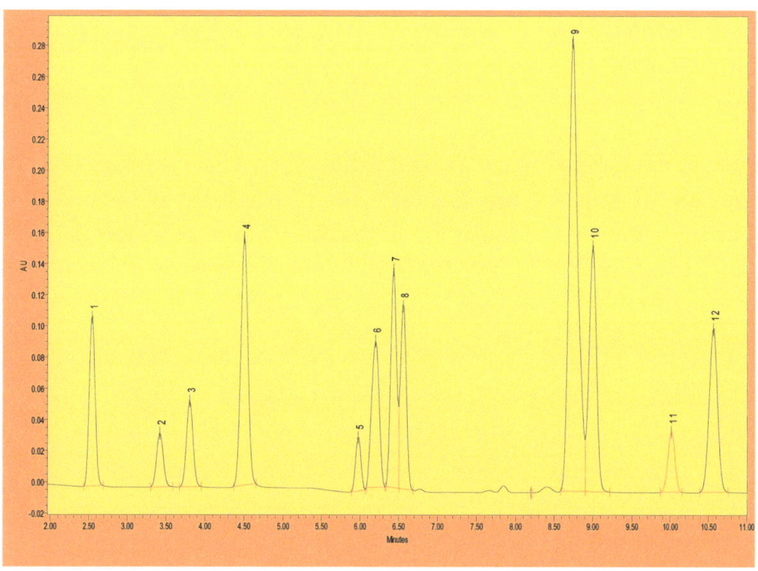

Figure 1. UPLC–UV chromatogram of a mixture of the following 12 pesticide reference standards, detected at 280 nm: acetamiprid (1), benomyl (2), flubendiamide (3), indoxacarb (4), carbosulfan (5), imidacloprid (6), methomyl (7), thiodicarb (8), spinosad (9), oxydemeton-methyl (10), difenoconazole (11), and mancozeb (12).

1 min, 0.2 mL/min flow rate, column and sample temperature were 45 and 25°C, respectively, injection volume was 1 μL and detection was done at 280 nm.

3.2. QuEChERS sample preparation method

As described, QuEChERS methodology [18, 19] have been adopted for the determination of 12 pesticide residues in chili. QuEChERS methodology have been devised in the year 2003 for the multiresidue analysis of pesticides in different matrices [20], and now it is universally accepted method [17]. In this procedure, extraction was performed with acetonitrile solvent initially and then partitioning step was carried out using salt mixture. A small amount of extract was further cleaned by using dispersive solid-phase extraction (d-SPE) method. Finally, extract was used for the determination of pesticide residues using UPLC. The advantages of this method include the large number of samples, and very low quantity of solvent and limited space are required [18, 21]. The acetonitrile has several advantages namely upon addition into salt, it will separate easily, good compatibility with d-SPE. The use of primary secondary amine removes acidic components, sugars and pigment molecules [18]. Another advantage is the removal of the waxes, lipids, and sugars during the freezing process. The pH of the extract will increases when it comes in contact with PSA [22]. This can be used as the stability of base-sensitive pesticides.

3.3. Method validation

Developed method has been validated after the optimization of the UPLC separation parameters. Limit of detections (LODs) were calculated using the signal to noise ratio by injecting 1 μL of dilute solutions.

3.3.1. Linearity

The calibration plot was constructed using the different concentrations namely 500 ppt, 1 ppb, 10 ppb, 1 ppm, and 10 ppm (**Figure. 2**) for checking the linearity of the method. Upto 10 ppm concentration, the response was linear for all the compounds, with correlation coefficient (R^2) values ranging from 0.9969 to 1.0000 (**Table 1**).

3.3.2. Accuracy and precision

Satisfactory results were found with recoveries between 85 and 100%. The relative standard deviation (RSD) was below 20%. The repeatability of the chromatographic method was determined by analyzing the chili samples spiked at different concentrations. The samples were injected 10 times with autosampler.

3.3.3. Limit of detection (LOD) and limit of quantification (LOQ)

For the blank sample of the chili, the limit of detection (LOD) of the compound can be measured using signal to noise ratio of 3 with obtained background noise. Then, for the limit of quantification (LOQ) of the method, S/N ratio was considered which was generally >10

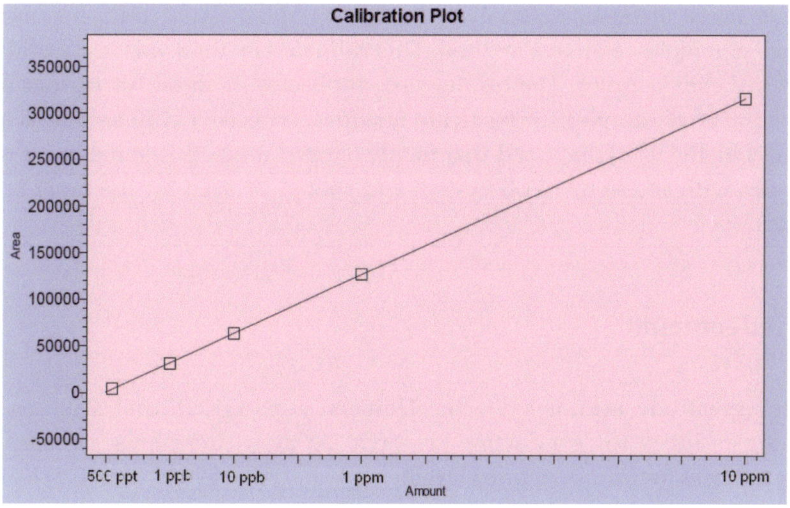

Figure 2. UPLC calibration plot of pesticide reference standards (500 ppt–10 ppm).

(**Table 1**). Effect of the matrix in the developed method was analyzed by comparing the standards in solvent with matrix-matched standards for five replicates. From the results obtained, it was evident that, no interfering peaks appeared and retention time (RT) of the tested analytes at spiked samples fully matched with those of standard samples. Each analyte molecule was eluted as separate symmetric peak.

3.3.4. *Analysis of pesticide residues in chili samples*

The validated method was employed for analysis of 30 samples collected from the different farmer's field of Haveri district, Karnataka, India. The optimized method was used for analysis of samples in triplicates. Results showed that most of the chili samples contained detectable pesticide residues (**Tables 2** and **3**). The residues of acetamiprid and thiodicarb were detected in three samples, whereas flubendiamide and mancozeb were detected in six samples, respectively, and carbosulfan and spinosad were detected in two and five number of samples, respectively (**Table 3**). The rest of the pesticides, that is, benomyl, indoxacarb, imidacloprid, methomyl, oxydemeton-methyl, and difenoconazole were not found in any of the samples. Out of the 30 chili samples, 19 samples did not contain any pesticide residues and 11 samples were found to be contaminated with residues with above MRLs.

4. Conclusion

Method has been developed with UPLC for the rapid detection and quantification of different pesticide residues in chili samples. The reliability of the method was checked by method validation in terms of linearity, precision, and accuracy in a range of 500 ppt–10 ppm, correlation coefficient (R^2) values were 0.9969. Average recoveries were more than 85–100%

for the wide range of pesticide analysis in chili samples. QuEChERS methodology has proved rapid and highly effective method. This validated method was successfully used for analysis of real chili samples. The results also emphasize the need for regular monitoring of a more number of samples for pesticide residues, especially chili sample which has to be exported. Finally, it is concluded that the developed method is suitable for routine use in laboratories with access to UPLC system and should be used for the rapid screening of chili samples.

Acknowledgements

The author's greatfully acknowledge the University of Agricultural Sciences, Dharwad for providing facilities for conducting research. Acknowledgements are also to ASIDE (Assistance to States for Infrastructure Development and Allied Activities) Govt. of India and Visvesvaraya Industrial Trade Centre (VITC) Govt. of Karnataka for providing grants for the establishment of Pesticide Residue Testing and Quality Analysis Laboratory.

Author details

Dinesh C Bilehal[1,2]*, Mahadev B. Chetti[4], Deepa G. T[3] and Mahadev C. Khethagoudar[2,3]

*Address all correspondence to: drbilehal@gmail.com

1 Department of Biomedical Science, Inje University, Gimhae, South Korea

2 Reva University, Bengaluru, Karnataka, India

3 University of Agricultural Sciences, Dharwad, Karnataka, India

4 Indian Council of Agricultural Research, New Delhi, India

References

[1] Mathur R, Dangi RS, Dass SC, Malhotra RC. The hottest chilli variety in India. Current Science. 2000;**79**(3):287-288

[2] CRN India. Available via http://www.crnindia.com.2010. [Accessed: 26-10-2010]

[3] FAOSTAT. FAO Statistical Database. http://www.fao.org.2008. [Accessed: 10-11-2010]

[4] Peter KV, Nybe EV, Mini Raj N. Available Technologies to Raise Yield. Surv. Ind. Agric. The Hindu Year Book. 2006. Chennai:82-86

[5] Lee IO, Lee KH, Pyo JH, Kim JH, Choi YJ, Lee YC. Anti-inflammatory effect of capsaicin in Helicobacter pylori-infected gastric epithelial cells. Helicobacter. 2007;**12**(5):510-517

[6] Deepa GT, Chetti MB, Khetagoudar MC, Adavirao GM. Influence of vacuum packaging on seed quality and mineral contents in chilli (*Capsicum annuum* L.). Journal of Food Science and Technology. 2013;**50**(1):153-158

[7] Ahuja KD, Robertson IK, Geraghty DP, Ball MJ. Effects of chili consumption on postprandial glucose, insulin, and energy metabolism. The American Journal of Clinical Nutrition. 2006;**84**(1):63-69

[8] Tomlin CDS (Ed.) The Pesticide Manual. 12th Ed. Surrey, UK: British Crop Protection Council; 2000

[9] Fernández, M, Picó Y, Mañs J. Analytical methods for pesticide residue determination in bee products. Journal of Food Protection. 2002;**65**(9):1502-1511

[10] Commission Directive 1999/39/EC of 6 May 1999. Amending Directive 96/5/EC on Processed Cereal-based Foods and Baby Foods for Infants and Young Children. Official Journal L124/8

[11] Picó Y, Font G, Moltó JC, Manes J. Pesticide residue determination in fruit and vegetables by liquid chromatography–Mass spectrometry. Journal of Chromatography A, 2000;**882**(1):153-173

[12] Torres CM, Picó Y, Manes J. Determination of pesticide residues in fruit and vegetables. Journal of Chromatography A. 1996;**754**(1):301-331

[13] Hajšlová J, Zrostlıkova J. Matrix effects in (ultra) trace analysis of pesticide residues in food and biotic matrices. Journal of Chromatography A. 2003:**1000**(1):181-197

[14] Yu K, Little D, Plumb R, Smith B. High-throughput quantification for a drug mixture in rat plasma–A comparison of Ultra Performance™ liquid chromatography/tandem mass spectrometry with high-performance liquid chromatography/tandem mass spectrometry. Rapid Communications in Mass Spectrometry. 2006;**20**(4):544-552

[15] Ayaz FA, Hayirlioglu-Ayaz S, Gruz J, Novak O, Strnad M. Separation, characterization, and quantitation of phenolic acids in a little-known blueberry (*Vaccinium arctostaphylos* L.) fruit by HPLC-MS. Journal of Agricultural and Food Chemistry. 2005;**53**(21):8116-8122

[16] Castellari M, Sartini E, Fabiani A, Arfelli G, Amati A. Analysis of wine phenolics by high-performance liquid chromatography using a monolithic type column. Journal of Chromatography A. 2002;**973**(1):221-227

[17] Anastassiades M, Lehotay SJ, Štajnbaher D, Schenck FJ. Fast and easy multiresidue method employing acetonitrile extraction/partitioning and "dispersive solid-phase extraction" for the determination of pesticide residues in produce. Journal of AOAC International. 2003;**86**(2):412-431

[18] Anastassiades M, Tasdelen B, Scherbaum E, Stajnbaher D Recent Developments in QuEChERS Methodology for Pesticide Multiresidue Analysis. Weinheim: Wiley-VCH; 2007

[19] European Committee for Standardization/Technical Committee 275 (Standards under development) Foods of Plant Origin. Determination of Pesticide Residues Using GC–MS and/or LC–MS (/MS) Following Acetonitrile Extraction/Partitioning and Cleanup by Dispersive SPE–QuEChERS Method. Brussels: European Committee for Standardization; 2007

[20] Lehotay SJ, Maštovská K, Yun SJ. Evaluation of two fast and easy methods for pesticide residue analysis in fatty food matrixes. Journal of AOAC International. 2005;**88**(2):630-638

[21] Lehotay SJ, Kok AD, Hiemstra M, Bodegraven PV. Validation of a fast and easy method for the determination of residues from 229 pesticides in fruits and vegetables using gas and liquid chromatography and mass spectrometric detection. Journal of AOAC International. 2005;**88**(2):595-614

[22] Lehotay SJ, Maštovská K, Lightfield AR. Use of buffering and other means to improve results of problematic pesticides in a fast and easy method for residue analysis of fruits and vegetables. Journal of AOAC International. 2005;**88**(2):615-629

Chapter 6

Advances in Distilled Beverages Authenticity and Quality Testing

Teodora Emilia Coldea, Elena Mudura and
Carmen Socaciu

Additional information is available at the end of the chapter

http://dx.doi.org/10.5772/intechopen.72041

Abstract

Given the advent of the consumers and producers demands, researches are focusing lately to develop innovative, cost-effective, progressively complex alcoholic beverages. As alcohol consumption has a heavy impact on social environment and health, fast and safe solutions for industrial application are needed. In this chapter, the recent advances in the field of alcoholic beverages authenticity and quality testing are summarised. Solutions for the online monitoring of the process of distilled beverages are offered and the recent methods for identification of raw material and process formed biomarkers of distilled beverages are presented.

Keywords: distilled beverages, authenticity, biomarkers

1. Introduction

Distilled beverages are important for consumers, producers and agricultural sector. Last decades presented us continuously changed requirements and descriptive practices for high level of consumer's protection with impact on the market transparency and fair competition. Both traditional methods and innovative technologies applied in distilled beverages production are focusing on their quality improvement.

The principal requirement set for an alcoholic beverage can be summarised as: are intended for human consumption, have specific sensory properties, with a minimum ethyl alcohol content of 15% v/v produced either by distillation with addition of flavourings, of naturally fermented products, or by addition of plant ethanol macerates, or by blending of flavourings, sugars, other sweetening products, or other agricultural origin products. Spirit drinks can also be produced by blending of different spirit drinks with ethyl alcohol of agricultural origin and other alcoholic or non-alcoholic beverages.

Generally, spirit drinks can be classified as with or without extract content. The presence of flavourings, sugars or other sweetening ingredients are forbidden in rum, whisky, vodka, grain spirit, wine spirit, grape or other fruit marc spirit, fruit, cider and perry spirits. No addition of other sources of ethanol of agricultural origin and no colour improvements by the addition of caramel are allowed in fruit spirits. In the category of distilled alcoholic beverages containing extract, we can specify plant macerates based spirit drinks, gin, aquavit, aniseed-flavoured spirit drinks, bitter, liqueur and mead nectar.

The European Commission prepared a list with the specific parameters and geographical indications of alcoholic beverages in countries across Europe. For example, only for fruit spirits are recognised 70 denominations, such as *Schwarzwälder Kirschwasser*, *sliwovitz*, *eau-de-vie*, *pálinka* and *țuica*.

The European countries have an old tradition in fruit growing and valorising in traditional distilled beverages. The traditional methods used to obtain distilled beverages involve the distillation of fermented plant material (fruits or cereals) in copper stills with open fire, maturing and conditioning in oak barrels, for at least 3 months. Usually distillation is repeated twice, such as in Romania or Hungary, frequently the ethyl alcohol content ranging over 50% v/v. No matter the production process applied, the flavour and taste of these distilled beverages should indicate the origin of the raw material used. Several stages, quality testing of the raw material, its preparation and fermentation are key factors determining the distilled beverages quality, with respect to their specific bouquet. Additionally, some of the major volatiles found especially in fruit distillates, such as methanol, furfural, isobutylic alcohol and acetaldehyde have toxic potential. Of main interest for consumer's health is the amount of methanol, which is the second compound found in fruit distillates after ethanol. It is usually ingested by consumers in low doses, but can create serious problems especially in countries with high unrecorded alcohol consumption [1].

Fruit spirits, very popular worldwide, are recognised in Eastern and Central Europe as a part of tradition heritage. Are considered as therapeutic agents since Middle Ages. The most famous fruit used is plum. In Europe, the most important countries producing fruit distillates are Poland, Slovakia, Hungary, Bulgaria, Romania and Czech Republic. Are used two categories of fruits—the one with stones (genus *Prunus*—plums, cherries, sour cherries, apricots, peaches etc.) and without stones (pears, apples and other berries). Each type of fruit give specific minor volatile compounds responsible for the aroma of the distillate—alcohols, aldehydes, esters, acids and volatile phenols [2]. The quality of a spirit is strongly related to the primary flavour given by the natural aroma of the fruit, which is influenced by the geographical origin of the fruit, method of cultivation, storage or harvest period. In this chapter, we describe recent advances in the field of alcoholic beverages authenticity and quality testing and indicate solutions for risk compounds decreasing during processing, that can be applied through beverages online monitoring.

2. Distilled beverages: health-related aspects

The beneficial effects of moderate alcohol consumption on dysfunctions of the cardiovascular system, such as coronary heart disease, associated myocardial infarction [3, 4], are diminished by the effects of alcohol to human health causing pancreatitis, diabetes, liver cirrhosis or pancreatic cancer [5–10].

It is well known the excessive alcohol consumption impact on social environment and health, recent studies showing the consequences of alcohol even to younger consumers [11–13]. European countries are confronted with a high level of alcohol consumption. With a range of 7.4% alcohol exposure of young people at the age of 15–29 represents the third major risk factor for human health causing premature death in the EU countries [14]. Actually, recent statistics place Romania as the first country in Europe for illicit alcohol consumption. Still, an exact amount of alcohol consumed cannot be given because, especially East European countries have their own tradition for homemade producing and consumption of distilled beverages [15–20].

According to epidemiological findings, the long-term consumption of alcoholic beverages is related to the occurrence of malignant tumours of the oral cavity, pharynx, larynx, oesophagus, liver, colorectal and female breast. Recent literature proved that ethanol carcinogenic mechanism is strongly linked with its transformations to acetaldehyde, already known for the carcinogenic activity.

Additionally, numerous outbreaks of alcohol poisoning, sometimes leading to fatal risk, were encountered especially with methanol, from illicit sources of alcoholic beverages. Some of the common symptoms of acute methanol poisoning are those related to hangover—headache, vertigo and vomiting, but can also cause severe abdominal pain, blurred vision or back pain. Still, its metabolites, formaldehyde and formic acid, are more harmful [21].

Furfural, a volatile compound derived from fruit carbohydrates, presents also toxicity to human organism consisting of inducing pain, sore throat, diarrhoea, vomiting and headache [16, 21]. It is formed during the improper conducted distillation, where commonly in homemade production of distilled beverages is used direct heating [2], creating harmful effects, caramel colour, with irreversible burnt-bitter taste. Acetaldehyde is formed during fermentation, by the ethanol dehydrogenation, and presents toxic effects associated with hangover-like symptoms such as nausea, sweating, rapid pulse, and headache and vomiting. It is also known as a carcinogenic compound [22, 23]. Distilled beverages, especially the homemade ones, may contain some amounts of heavy metals—such as lead or copper. Lead occurrence in distilled beverages comes from the pesticides used in agriculture and remained in ground water. When the alcohol content of traditional beverages is adjusted, usually no water analyses are made, and, as a consequence, water can become a harmful source of chemical pollution. Copper provenience in distilled beverages, namely in traditional ones, is from distilling installation or, as in the case of lead, from the pesticides used in agriculture [24]. As a positive effect, copper is key determinant in the improvement of sensory characteristics of many alcoholic beverages [25].

Recently, European Food Safety Authority (EFSA) focused on ethyl carbamate content (a derivative from hydrogen cyanide (HCN) during fruit distillates processing). Even though European Regulation (110/2008) established limits for contaminants in alcoholic beverages, sometimes the imposed limits are still exceeded. The literature presents data on advanced techniques applied for determination of HCN, formed during fermentation process, from fresh fruit, fruit juice and kernel by [26].

3. Quality parameters of distilled beverages

The most important quality parameters, which are related to the safety of distilled beverage are ethanol content, esters, aldehydes, higher alcohols, methanol, furfural and HCN. These are also specified in the EU regulation for each type of product.

Depending on the product's type, different values for ethanol content are imposed by EU regulation (min. 96% v/v in ethanol of agricultural origin; min. 37.5% v/v in vodka, fruit spirits, rum and wine spirit; min. 40% v/v in whisky).

Maximum level of methanol in ethanol of agricultural origin is 30 mg/100 mL p.a. High concentration of methanol in fruit distillates is directly related to the quantity of pectins present in fruits, which are methoxylated during fruit riping. Methanol forms when pectic substances hydrolyse under the influence of some pectolytic enzymes. As a result of pectases action, demethylation can occur releasing methanol together with pectic acid and pectol. One of the objectives of the second distillation is the concentration of methanol in overhead fraction, as to be removed and reduce its content in final distillate to a concentration in accordance with the maximum admissible levels (1200 mg/100 mL pure alc. in fruit distillates, 200 mg/100 mL p.a. in wine spirit, 1000 mg/100 mL p/a/ in cider spirit and 30 mg/100 mL p.a in ethyl alcohol of agricultural origin).

Furfural is a chemical compound that, in small amounts, contributes to the aroma and bouquet of fruit distillates, and is not allowed in ethanol of agricultural origin due to its possible toxic effect. Its health harmful effects are skin, eyes and respiratory tract irritation, headache, taste loss, skin allergies, respiratory difficulties, vomiting, thirst sensation and long exposure can affect the central nervous system, liver or blood.

The mean intake of ethyl carbamate from food is approximately 15 ng/kg bw per day, excluding the levels that come from alcoholic beverages. High levels of ethyl carbamate (ranging between 0.01 and 12 mg/L) can be found in distilled spirits, mainly in stone fruit spirits [20], depending on their origin [27]. As more than 80% of ethyl carbamate is formed, the next 4 h after distillation ends, it is important to avoid its accumulation by applying few measures: chemical elimination of cyanide in fermented juices and after distillation and the replacing of copper condensers with stainless steel ones or proper separation of heads, as HCN has a low boiling point—25.7°C [28].

Total volatile substances are a quality indicator for alcoholic beverages. The more amounts of volatile substances in fruit distillate (min. 200 mg/100 mL p.a.) or rum (min. 225 mg/100 mL p.a.) increases their quality.

http://dx.doi.org/10.5772/intechopen.72041

Along with common analytical methods, it is imposed the necessity of some rapid, eco-friendly and cost-effective simplified alternative for alcoholic beverages assessment with applicability for both research and authorities.

4. Extraction methods applied in distilled beverages analysis

With the advent of the modern scientific revolution and the development of chemistry, alcoholic beverage sector becomes progressively complex, new ideas passing very fast from a research theme to a final market product. Many researches were conducted in this field and provide solutions with fast industrial applicability.

Given the large number of existed and daily worldwide innovated beverages, considerable progress in terms of extraction methods and analytical techniques have been made especially in the last decade. Different extraction methods such as liquid-liquid extraction (LLE), headspace solid-phase microextraction (HS-SPME), stir bar sorptive extraction (SBSE), gas-chromatography and ultra high performance liquid chromatography (UPLC) have been tested and discussed. **Table 1** summarises some of the techniques generally used for the compound extractions and the analytical methods applied for the characterisation of distilled beverages.

Type of distilled beverage	Investigated compound	Sample preparation	Analytical techniques	References
Non-aged fruit spirit	Volatile compounds	HS-SPME	GC-MS	[15, 29]
	Volatile compounds	–	GC-MS	[30, 31]
	Volatile compounds	LLE	GC-MS	[2]
	Volatile compounds	LLE	GC-O	[32]
	Volatile compounds	–	GC-FID	[16, 30, 31]
	Sensory parameters	–	Sensory analysis	[33]
	Volatile compounds	–	Ethanol, methanol (FTIR)	[2]
	Phenols	Solvent extraction	Total phenolics (UV-VIS) Antioxidant activity (DPPH, FRAP) Phenolics (HPLC)	[34]
	Metal content	–	Heavy metals (Atomic absorbtion spectrophometry)	[31, 35]
Vodka	Phenols	Solvent extraction	Total phenolics (UV-VIS) Antioxidant activity (DPPH, FRAP) Phenolics (HPLC)	[34]

Type of distilled beverage	Investigated compound	Sample preparation	Analytical techniques	References
Rum	Phenols	Solvent extraction	Total phenolics (UV-VIS) Antioxidant activity (DPPH, FRAP) Phenolics (HPLC)	[34]
Bitter	Phenols	Solvent extraction	Total phenolics (UV-VIS) Antioxidant activity (DPPH, FRAP) Phenolics (HPLC)	[34]
Bitter liqueur	Phenols	Solvent extraction	Total phenolics (UV-VIS) Antioxidant activity (DPPH, FRAP) Phenolics (HPLC)	[34]
Fruit liqueur	Volatile compounds	HS-SPME	GC-MS	[31]
	Phenols	Solvent extraction	Total phenolics (UV-VIS) Antioxidant activity (DPPH, FRAP) Phenolics (HPLC)	[34]
Aged distillate	Phenols	Solvent extraction	Total phenolics (UV-VIS)	[34, 36–39]
	Phenols	Solvent extraction	Phenolics (HPLC)	[33, 34, 36, 37, 40–44]
	Phenols	Solvent extraction	Phenolics (UPLC)	[45]
	Phenols	Solvent extraction	Antioxidant activity (DPPH, FRAP)	[33, 34]
			Antioxidant activity (ABTS)	[39]
	Colour parameters	–	Colour intensity and hue (UV-VIS)	[36]
	Sensory parameters	–	Sensory analysis	[36, 37, 46]
	Volatile compounds	LLE	GC-MS GC-FID	[46, 47]
		SPME	GC-MS	[44]
		–	GC-O	[48]
	Volatile compounds	Stir bar sorptive extraction (SBSE)	GC-MS	[49, 50]

Table 1. Some techniques used for the extraction of distilled beverages interest compounds and the analytical methods applied for their quality, safety and authenticity assessment.

5. Authenticity biomarkers

5.1. Raw material biomarkers

The quality of final beverage is strongly dependent of the quality of raw material used, variety, harvesting methods applied and storage conditions. In this context, several studies focused the chemical characteristics, volatile and microbiological aspects of vegetal matrices used in production of distilled beverages [20, 51–54]. The applied techniques in volatile compounds assessment are infrared spectroscopy [55], gas-chromatography coupled with mass-spectrometry [51, 52, 56, 57] and gas-chromatography with flame ionisation detector [53, 58].

Liquid chromatography was applied for identification and quantification of phenolic compounds in apple [53, 59, 60], nectarin, peaches and plums [61].

Free fatty acids from fruits are considered main responsible for the beverages aroma studies found [51, 62–64]. Fatty acid esters are contributors of fruit distillates flavour giving the fruity and flowery notes [2].

5.2. Biomarkers formed during fermentation process

Fermentation represents one of the most important factors for the quality and complexity of distilled beverages. Despite common laboratory methods used to test de-fermented marc quality, recent studies applied advanced methods. The quality of raw material subjected to fermentation, yeast specie and the inoculum amount, type and hygiene of the vessels [65], temperature and duration of the process are parameters intensively tested in the past years [20, 26, 28, 66, 67]. Yeast species used in fermentation have an important role in defining the final bouquet of the beverage, a gas-chromatography analysis proved [68]. Molecular techniques were used for distinguishing different types of microorganisms involved in the fermentation of cachaça [69] with impact on their volatile profile [70].

Ethanol, the first fermentation metabolite, along with high alcohols, aldehydes and fatty acids with their esters, are responsible for the flavour formation. In the case of traditional fruit, distillates are not involved in selected yeast cultures. The fermentation is carried out spontaneously, in the presence of *Kloeckera apiculata* yeast species present on fruit peels surface. Yeast microbiota species responsible for alcoholic fermentation of fruits depends especially by the region of raw material, the fermentation process applied, type of the final beverage, initial cellular concentration, temperature, pH, sulphur dioxide content and ethanol concentration. Fermentation microbiota is formed of *Saccharomyces cerevisiae* and other spontaneous species, which should not exceed 10^6–10^9 CFU/mL, in order not to inhibit the *S. cerevisiae* biological activity. During the fermentation, *S. cerevisiae* will dominate the must microbiota due to their resistance to formed ethanol. Along with *S. cerevisiae*, which represents the majority of must yeast when must ethanol concentration exceeds 5% v/v, fermentation mibrobiota is formed by *K. apiculata* (representing 98% of the viable microbiota before must exceeded 5% v/v ethanol concentration), *Rhodotorula mucilaginosa*, *R. graminis* and *Aureobasidium* sp. [66]. These spontaneous yeast species can demonstrate the authenticity of a specific fermented beverage. Contrarily, must fermentation in the presence of selected strains of *S. cerevisiae*, provide more complex volatile profile of the final distillate [71]. Even more, were found ethyl decanoate and ethyl-2-*trans*-4-*cis*-decadienoate as authenticity biomarkers for pear distillates [72]. The higher number of viable fermentation yeast contributes to more esters formation. During fermentation, are formed also eugenol (that comes from phenol or ferulic acid), acetic acid and other ethyl esters [73].

Acetaldehyde, along with ethyl hexanoate, octanoate, decanoate and dodecanoate esters, which after distillation and ageing increase, are also compounds formed during fermentation [74].

Acetaldehyde, ethyl acetate and amyl alcohols, are the main responsible of distilled beverages aroma, formed during fermentation, their amounts influencing the quality of distillates [75].

5.3. Biomarkers formed during distillation process

The influence of distillation on the distillate bouquet is decisive. Distillation influences both the products quality and safety. Unwanted fractions, considered as toxic for human organism, are separated in this step. Separation depends on the distillation procedure (simple, double with or without rectification), temperature, duration, compounds separation grade and the fermented marc quality. The right moment for distillation is very important, because influences also the distillation yield, and should be performed immediately after the fermentation ends. The way distillation is conducted strongly influences the distillate quality. Marc heating should be made gradually and uniform, in order to avoid the burning taste and smell.

By distillation, components of a homogenous liquid are separated based on their constituents boiling temperatures and vapour pressures. The most volatile components have higher vapour pressures, so it will concentrate in the vapour phase, and will form the condensate. The low volatile compounds will remain in the residual liquid. The volatile compounds formed in fermentation are extracted by distillation. Usually, are used stainless steel installation, but the traditional method includes copper alembic. Copper has an important role as reaction catalyser and formation of aromatic substances. Firstly, marc is introduced in boiling kettle, from where will be obtained the first distillate, which is subjected to a new distillation. Redistillation aims the separation of heads and tails fractions. After the second distillation, ethanol content will be range 50% v/v.

The principal compounds resulted from distillation are water, ethanol and hundreds of volatile compounds, contributors to distillate flavour. During distillation three main fractions are collected: heads, middle fraction and tails (fusel oils).

Acetaldehyde, propionaldehyde, methanol and some esters (ethyl acetate, methyl acetate and acetal) form heads fraction. They are highly toxic compounds, present unpleasant smell and taste and are separated after the first distillation. Middle fraction (ethanol content interval between 20 and 72% v/v) represents 30–45% of the entire distillate, and it contains ethanol, higher alcohols, acetals, which gives distillate specific aroma. Tails represent a high amount of volatile compounds, with high boiling point and water solubility. This fraction (10–15% of the entire amount of distillate) includes furfural, acetic acid, ethyl lactate, fatty acids with high molecular mass, fatty acids esters, volatile acids (propionic, butyric, isovalerianic and capronic) and higher alcohols (amylic, isoamylic and hexylic), and is also a strongly toxic fraction, with unpleasant sensorial properties, which is the reason for need of their separation. Heads are subjected to a new distillation, while tails are added to the marc, in order to perform the second distillation.

Many studies were performed for the identification of the different types of distilled beverages composition. A study performed in Thailand on rice distillate found volatiles identified by GC-MS have the strong odorant capacity: ethyl acetate, ethyl butyrate, ethyl decanoate, acetaldehyde, ethyl laureate, ethyl caprilate, 2-phenethyl acetate, 1-hexanol, isoamylic alcohol and 2-furaldehyde [76]. The volatile phenols identified were 4-ethylguaiacol, eugenol and 4-ethylphenol, responsible of strong floral and spicy notes, and they should be only in moderate amounts when compared to the other compounds. Grape marc distillate contains higher alcohols (1-hexanol, *trans*-3-hexen-1-ol, *cis*-3-hexen-1-ol and *trans*-2-hexen-1-ol), fusel alcohols (1-propanol, 2-butanol, isobutanol, 1-butanol, isoamyl alcohol, 1-pentanol, 1-octen-3-ol, 1-heptanol, 1-octen-4-ol, 1-octanol, 1-nonanol, 1-decanol, benzyl alcohol and 2-phenylethanol), fatty acids (isobutyric, butyric, pentanoic, hexanoic, heptanoic, octanoic, decanoic, dodecanoic, tetradecanoic and hexadecanoic acids), fatty acids esters (ethyl propanoate, ethyl isobutyrate, ethyl butyrate, ethyl 2-methyl butyrate, ethyl 3-methyl butyrate, ethyl pentanoate, ethyl hexanoate, ethyl heptanoate, ethyl octanoate, ethyl nonanoate, ethyl decanoate, ethyl dodecanoate, ethyl tridecanoate, ethyl tetradecanoate, ethyl hexadecanoate, ethyl octadecanoate, ethyl linoleate and ethyl linolenate), acetic esters (ethyl acetate, propyl acetate, isobutyl acetate, isoamyl acetate and hexyl acetate), hydroxi and dicarboxy acids esters (ethyl lactate and isoamylic lactate) and other esters [77].

Moderate presence of ethyl acetate is considerate as a positive aspect due to its aroma flavour. Though, acetals in large amounts indicate a possible microbial contamination. Acetaldehyde brings hazel, cherry and overripe apple flavours.

Furfural is formed during distillation by acid hydrolysis, by fermented pentoses heating or by Maillard reactions, especially by direct heating in copper alembic. The distillation method applied influences especially the amount of esters in the final distillate, mainly considering the increasing of ethyl decanoate and ethyl palmitate [78]. When distillation columns are used, a significant increase of esters is observed, 20% more higher alcohols, less aldehyde (a decrease with 40%) and 10% less methanol, when compared to alembic distillation. Higher concentrations of methyl and ethyl acetates indicate the incorrect fractions separation (heads) during the distillation process [65]. Lactate ethyl (found in tails fraction), with unpleasant aroma, is considered a marker of the long fermented marc storage before distillation, when unwanted malolactic fermentation takes place. Fatty acids esters (caproic, caprilic, capric and lauric), with a fruity floral flavour, are formed during fermentation. Ethyl caproate presents a banana flavour, ethyl caprylate is less perfumed, and giving a grape oil flavour, ethyl caprate is less intense, while ethyl laurate gives a candle wax smell. Another source for fatty acids esters formation can be explained by the yeasts thermal degradation and autolysis during distillation.

Methanol is separated in heads and tails. Still, a small amount of methanol is separated also in middle fraction [79], depending on the fruits processing method (crushing or pressing), storage duration, fruits initial composition, pH, fermentation and distillation temperatures [65].

Higher alcohols, isoamyl alcohol and 1-propanol in the greatest amount, are known as fusel oils, having a higher boiling point than ethanol. They present a strong flavour and are retrieved in large amounts in distillate beverages, depending on the raw material and yeasts species used during fermentation, as these compounds are derived from the sugars and amino acids metabolised by yeasts [79].

Aldehydes, the most volatile compounds in distilled beverages, are formed during fermentation and are considered the main compounds resulted from the biochemical reaction when yeasts use amino acids and sugars. Acetaldehyde is the most abundant (90% of total aldehydes) and it is accumulated in heads fraction [79]. Acrolein is another important aldehyde present in heads fraction of the distillate, has a strong spicy aroma and is produced by bacteria from glycerol.

5.4. Aged distillates biomarkers

All distilled beverages need a maturation period before consumption. During maturation, a physical structuring of the ethanol and water molecules is produced and as a result, distillate becomes smoother and less pungent [80].

Biochemical process during distilled beverages ageing depends on the temperature, oxygen and the chemical composition of the ageing or maturation materials. The storage materials are very important for the distillate safety. When plastic material is used, unwanted high extraction yield of phthalate compounds was observed especially in illegally produced strong alcoholic beverages [81].

Distillates ageing can be performed in different wooden barrels or by the addition of different wood powders and other wood fragments.

Ever since OIV approved the use of wood alternatives for barrels, different methods are applied to shorten the ageing period of alcoholic beverages and enhance their phenolic and flavour profile [82]. Recently, more attention is given to the use of wooden fragments and powders for the rapid inducing ageing character of brandies [37, 46, 83–85], with reducing the lasting period between days to some weeks. The cost and difficulty of wooden barrels handling, guided actors in beverage industry to these less expensive alternatives.

Free phenolic compounds from oak wood, in contact with oxygen produce quinones. Surface wood phenols, partially transformed by quinones, dissolved in distillate and, along with quinones oxidise distillate compounds with the formation of other phenolic compounds. Furfural can be formed also during ageing process of distillate. It results by pentose oxidation by wood extraction and by wood sugar residues.

Distillate storage in wooden barrels is producing a high ethanolic extraction of wood components in distillate, especially referring to gallic acid and quercetin. The action of tanase on gallic acid produces an oxidation reaction, resulting a gold-yellow colour of the distillate. Simultaneous are produced reactions of oxidation of ethanol, higher alcohols and aldehydes with formation of acids, which react with alcohols forming esters with specific flavour.

During ageing, as interaction between wood compounds (hemicelluloses, lignin, phenolic compounds, cellulose) and distillate, are produced physical (volume decreasing, alcohol content loss, colour intensity increasing, specific mass and extract increasing) and chemical transformations (oxido-reduction processes, pH modifications, new ester, acetals and aldehydes formations). Some of the substances absorbed from wood in distillate are colour, pectic and mineral compounds, amino acids and sugars. The most important are phenolic compounds with impact on both sensory properties of the distillate and on its antioxidant activities. Depending on wood species used in cooperage, different wood-related phenols are extracted in distilled beverages (**Table 2**). Despite oak (*Quercus robur* L.), the most frequently researched woods in cooperage are chestnut (*Castanea sativa* Miller), cherry tree (*Prunus avium* L.), walnut (*Juglans regia* L.), acacia

(*Robinia pseudacacia* L.), mulberry (*Morus alba* and *Morus nigra*), ash (*Fraxinus excelsior* L. and *Fraxinus Americana* L.), beech (*Fagus sylvatica* L.), alder (*Alnus glutinosa* L.) and lime (*Tilia cordata* Miller) [33, 86–89], although only oak and chestnut are approved by OIV for (wine) ageing [89].

5.5. Rapid authenticity testing of distilled beverages

Ethyl alcohol used for the production of spirit drinks should be of exclusively agricultural origin (EUR Lex 110/2008). A key quality parameter is the maturation and ageing periods, respectively.

Phenolic compound	Oak	Chestnut	Cherry	Mulberry	Walnut	References
Gallic acid	*	*	*		*	[82, 86, 88, 90, 91]
Vanillic acid	*	*	*			[82, 86, 88, 91, 92]
Syringic acid	*	*	*			[82, 86, 88, 91]
Vanillin	*	*	*	*		[38, 82, 83, 86, 88, 90, 92]
Ellagic acid	*	*	*		*	[82, 86, 88, 91]
Trans caffeic acid	*					[91]
Trans cafftaric acid	*					[91]
5-OH-methyl-furfural	*					[82, 83]
Furfural	*	*				[38, 82, 92]
Syringaldehyde	*	*	*			[38, 82, 86, 88, 92]
Coniferaldehyde	*	*	*			[38, 82, 86, 88]
Sinapaldehyde	*	*	*			[82, 86, 88]
Eugenol	*	*	*	*		[38, 83, 90, 92]
Methoxyeugenol	*					[90]
Bnezene derivatives				*		[90]
Trimethoxyphenol			*			[90]
Benzaldehyde derivatives			*			[90]
Protocatechuic acid	*	*	*			[86, 88, 91]
p-Hydroxybenzoic acid	*		*			[88, 91]
Ferulic acid	*	*	*			[86, 88]
Protocatechuic aldehyde	*	*	*	*		[86, 88]
Vanillic aldehyde		*				[86]
Coumarin		*				[86]
Scopoletin	*	*				[86, 88]
Vescalagin	*	*	*			[86, 88]
Castalagin	*	*	*			[86, 88]

Phenolic compound	Oak	Chestnut	Cherry	Mulberry	Walnut	References
Acutissimin		*				[86]
Glucopyranose derivatives		*				[86]
Ellagic acid deoxyhexose		*				[86]
Ellagic acid dimer dehydrated		*				[86]
Valoneic acid dilactone		*				[86]
p-Coumaric acid	*	*	*			[86, 88]
cis p-Coumaric	*					[91]
trans p-Coumaric	*					[91]
cis Coutaric	*					[91]
trans Coutaric	*					[91]
cis p-Coumaric derivative	*					[91]
trans p-Coumaric derivative	*					[91]
Methyl vanillate			*			[86]
Methyl syringate			*			[86]
Benzoic acid			*			[86]
Flavan-3-ols			*			[86]
(+)-Catechin	*		*			[86]
(−)-Epicatechin	*					[91]
B-type procyanidin dimer			*			[86]
B-type procyanidin trimer			*			[86]
Naringenin			*			[86]
Isosakuranetin			*			[86]
Eriodictyol			*			[86]
Aromadendrin			*			[86]
Taxifolin			*	*		[86]
Guaiacol	*					[83]
trans-Oaklactone	*					[83]
cis-Oaklactone	*					[83]
o-Cresol	*					[83]
Ethyl guaiacol	*			*		[38, 83]
p-Cresol	*					[83]
2,6-Dimethoxy-phenol	*					[83]
Myricetin 3-O-glucoside	*					[91]
Quercetin 3-O-glucuronide	*					[91]
Quercetin 3-O-galactoside	*					[91]

Phenolic compound	Oak	Chestnut	Cherry	Mulberry	Walnut	References
Myricetin	*					[91]
Quercetin	*					[91]
Resveratrol trans 3-O-glucoside	*					[91]
Resveratrol cis 3-O-glucoside	*					[91]
Resveratrol trans	*					[91]
Resveratrol cis	*					[91]
Resveratrol				*		[91]
Oxyresveratrol				*		[91]
Dihydroxyresveratrol				*		[91]
Viniferin	*					[91]
2,3-Hydroxy-1-guayacyl-Propan-1-one	*					[91]
Methyl gallate	*					[91]
Tyrosol	*					[91]
Ethyl gallate	*					[91]
Tryptophol	*					[91]
4-Methylguaiacol				*		[38]
Sinapic acid	*	*	*			[88]
Caffeic acid	*	*	*			[88]
Roburin A	*	*				[88]
Roburin B	*	*				[88]
Roburin C	*	*				[88]
Roburin D	*	*				[88]
Roburin E	*	*				[88]
Grandinin	*	*				[88]

Table 2. Phenolic compounds identified in wood extracts as mentioned in different references.

The addition of flavourings, sugars or other sweetening products is forbidden for distilled beverages such as rum, whisky, fruit distillates or wine brandy. The addition of caramel in fruit distillates is not allowed, and in whiskey is allowed only the plain caramel (for colouring) [93], being considered as counterfeit.

Advanced techniques for laboratory analysis of the alcoholic beverages are the chromatographic ones. The expensive reference methods tend to be replaced by simpler ones, non-destructive and easy to handle. The Fourier Transform Infrared Spectroscopy (FTIR) technique, in combination with chemometrics is a fast and reproducible way to identify the authenticity and adulteration of beverages [19, 94]. As for wine, vinegar or olive oil, distilled beverages have also geographical indication denominations. Recent studies focused the discrimination of distilled beverages based on their raw material and geographical region [2, 95–97]. Due to the degradation of methoxylated pectins found in pulp, fruit distillates will content higher amounts of methanol in comparison with other matrices (cereal ethyl alcohol). Indirectly, methanol can be considered an authenticity marker of a natural fruit brandy, indicating the origin of raw material used [16].

Author details

Teodora Emilia Coldea[1]*, Elena Mudura[1] and Carmen Socaciu[2]

*Address all correspondence to: teodora.coldea@usamvcluj.ro

1 Department of Food Technology, Faculty of Food Science and Technology, University of Agricultural Sciences and Veterinary Medicine, Cluj-Napoca, Romania

2 Department of Food Science, Faculty of Food Science and Technology, University of Agricultural Sciences and Veterinary Medicine, Cluj-Napoca, Romania

References

[1] Levy P, Hexdall A, Gordon P, Boeriu C, Heller M, Nelson L. Methanol contamination of Romanian home-distilled alcohol. Journal of Toxicology. Clinical Toxicology. 2003;**41**:23-28. DOI: 10.1081/CLT-120018267

[2] Coldea TE, Socaciu C, Moldovan Z, Mudura E. Minor volatile compounds in traditional homemade fruit brandies from Transylvania-Romania, as determined by GC-MS analysis. Notulae Botanicae Horti Agrobotanici Cluj-Napoca. 2014;**42**:530-537 DOI: 10.15835/nbha4229607

[3] Brien SE, Ronksley PE, Turner BJ, Mukamal KJ, Ghali WA. Effect of alcohol consumption on biological markers associated with risk of coronary heart disease: Systematic review and meta-analysis of interventional studies. BMJ Open. 2011;**342**:d636. DOI: 10.1136/bmj.d636

[4] Koppes LLJ, Dekker JM, Hendriks HFJ, Bouter LM, Heine RJ. Moderate alcohol consumption lowers the risk of type 2 diabetes: A meta-analysis of prospective observational studies. Diabetes Care. 2005;**28**:719-725. DOI: 10.2337/diacare.28.3.719

[5] Witt H, Apte MV, Keim V, Wilson JS. Chronic pancreatitis: Challenges and advances in pathogenesis, genetics, diagnosis, and therapy. Gastroenterology. 2007;**132**:1557-1573. DOI: 10.1053/j.gastro.2007.03.001

[6] Yadav D, Whitcomb DC. The role of alcohol and smoking in pancreatitis. Nature Reviews Gastroenterology and Hepatology. 2010;**7**:131-145. DOI: 10.1038/nrgastro.2010.6

[7] Deng X, Wang L, Elm MS, Gabazadeh D, Diorio GJ, Eagon PK, Whitcomb DC. Chronic alcohol consumption accelerates fibrosis in response to cerulein-induced pancreatitis in rats. American Journal of Pathology. 2005;**166**:93-106. DOI: 10.1016/S0002-9440(10)62235-3

[8] Go VLW, Gukovskaya A, Pandol SJ. Alcohol and pancreatic cancer. Alcohol. 2005;**35**:205-211. DOI: 10.1016/j.alcohol.2005.03.010

[9] Hassan MM, Bondy ML, Wolff RA, Abbruzzese JL, Vauthey JN, Pisters PW, Evans DB, Khan R, Chou TH, Lenzi R, Jiao L, Li D. Risk factors for pancreatic cancer: Case-control study. American Journal of Gastroenterology. 2007;**102**:2696-2707. DOI: 10.1111/j.1572-0241.2007.01510.x

[10] Lachenmeier DW, Monakhova YB, Rehm J. Influence of unrecorded alcohol consumption on liver cirrhosis mortality. World Journal of Gastroenterology. 2014;**20**:7217-7222

[11] Salanţă LC, Tofană M, Pop C, Pop A, Coldea T, Mudura E. Beverage alcohol choice among university students: Perception, consumption and preferences. Bulletin UASVM Food Science and Technology. 2017;**74**:23-30. DOI: 10.15835/buasvmcn-fst:12633

[12] McCambridge J, McAlaney J, Rowe R. Adult consequences of late adolescent alcohol consumption: A systematic review of cohort studies. PLoS Medicine. 2011;**8**:e1000413. DOI: 10.1371/journal.pmed.1000413

[13] Berkey CS, Willett WC, Frazier AL, Rosner B, Tamimi RM, Rockett HRH, Colditz GA. Prospective study of adolescent alcohol consumption and risk of benign breast disease in young women. Pediatrics. 2010;**125**:e1081-e1087. DOI: 10.1542/peds.2009-2347.

[14] Salanţă LC, Tofană M, Mudura E, Pop C, Pop A, Coldea T. The alcoholic beverage consumption preference of university students: A preliminary Romanian case study. Bulletin UASVM Food Science and Technology. 2016;**73**:33-39. DOI: 10.15835/buasvmcn-fst:11984

[15] Bajer T, Bajerová P, Surmová S, Kremr D, Ventura K, Eisner A. Chemical profiling of volatile compounds of various home-made fruit spirits using headspace solid-phase microextraction. Journal of the Institute of Brewing. 2017;**123**:105-112. DOI: 10.1002/jib.386

[16] Rusu Coldea TE, Socaciu C, Pârv M, Vodnar D. Gas-chromatographic analysis of major volatile compounds found in traditional fruit brandies from Transylvania, Romania. Notulae Botanicae Horti Agrobotanici Cluj-Napoca. 2011;**39**:109-116. DOI: 10.15835/nbha3926053

[17] Rusu Coldea TE, Socaciu C, Fetea F, Ranga F, Pârlog R. Phenolic derivatives as authenticity markers of traditional homemade brandies from different counties of Transylvania, using UV-VIS and HPLC analysis. Bulletin UASVM Agriculture. 2011;**68**:518-528. DOI: 10.15835/buasvmcn-agr:6610

[18] Pecić S, Veljović M, Despotović S, Leskošek-Ćukalović I, Jadranin M, Tešević V, Nikšić M, Nikićević N. Effect of maturation conditions on sensory and antioxidant properties of old Serbian plum brandies. European Food Research and Technology. 2012;**235**:479-487. DOI: 10.1007/s00217-012-1775-y

[19] Coldea TE, Socaciu C, Fetea F, Ranga F, Pop RM, Florea M. Rapid quantitative analysis of ethanol and prediction of methanol content in traditional fruit brandies from Romania, using FTIR spectroscopy and chemometrics. Notulae Botanicae Horti Agrobotanici Cluj-Napoca. 2013;**41**:143-149. DOI: 10.15835/nbha4119000

[20] Śliwińska M, Wiśniewska P, Dymerski T, Wardencki W, Namieśnik J. The flavour of fruit spirits and fruit liqueurs: A review. Flavour and Fragrance Journal. 2015;**30**:197-207. DOI: 10.1002/ffj.3237

[21] Rohsenow DJ, Howland J. The role of beverage congeners in hangover and other residual effects of alcohol intoxication: A review. Current Drug Abuse Reviews. 2010;**3**:76-79. DOI: 10.2174/1874473711003020076

[22] Pflaum T, Hausler T, Baumung C, Ackermann S, Kuballa T, Rehm J, Lachenmeier DW. Carcinogenic compounds in alcoholic beverages: An update. Archives of Toxicology. 2016;**90**:2349-2367. DOI: 10.1007/s00204-016-1770-3

[23] Lachenmeier DW, Monakhova YB. Short-term salivary acetaldehyde increase due to direct exposure to alcoholic beverages as an additional cancer risk factor beyond ethanol metabolism. Journal of Experimental & Clinical Cancer Research. 2011;**30**:3-9. DOI: 10.1186/1756-9966-30-3

[24] Rusu Coldea TE, Socaciu C, Părău C, Mocan AA. Quality and safety analysis for some traditional homemade fruit distillates from Transylvania (North West Romania). Bulletin UASVM Agriculture. 2010;**67**:395-403. DOI: 10.15835/buasvmcn-agr:5155

[25] Bonić M, Tešević V, Nikićević N, Cvejić J, Milosavljević S, Vajs V, Mandić B, Urošević I, Veličković M, Jovanić S. The contents of heavy metals in Serbian old plum brandies. Journal of the Serbian Chemical Society. 2013;**78**:933-945. DOI: 10.2298/JSC121106016B

[26] Balcerek M, Szopa J. Ethanol biosynthesis and hydrocyanic acid liberation during fruit mashes fermentation. Czech Journal of Food Sciences. 2012;**30**(2):144-152

[27] Weber JV, Sharypov VI. Ethyl carbamate in foods and beverages: A review. Environmental Chemistry Letters. 2009;**7**:233-247. DOI: 10.1007/s10311-008-0168-8

[28] Spaho N. Distillation techniques in the fruit spirits production. In: Mendes M, editor. Distillation – Innovative Applications and Modeling. InTech; 2017. pp. 129-152. DOI: 10.5772/66774

[29] Zhao Y, Xu Y, Li J, Fan W, Jiang W. Profile of volatile compounds in 11 brandies by headspace solid-phase microextraction followed by gas chromatography-mass spectrometry. Journal of Food Science. 2009;**74**:c90-c99. DOI: 10.1111/j.1750-3841.2008.01029.x

[30] Versini G, Franco MA, Moser S, Barchetti P, Manca G. Characterisation of apple distillates from native varieties of Sardinia island and comparison with other Italian products. Food Chemistry. 2009;**113**:1176-1183. DOI: 10.1016/j.foodchem.2008.08.003

[31] Galego LR, Da Silva JP, Almeida VR, Bronze MR, Boas LV. Preparation of novel distinct highly aromatic liquors using fruit distillates. International Journal of Food Science and Technology. 2011;**46**:67-73. DOI: 10.1111/j.1365-2621.2010.02452.x

[32] Malfondet N, Gourrat K, Brunerie P, Le Quéré JL. Aroma characterization of freshly-distilled French brandies; their specificity and variability within a limited geographic area. Flavour and Fragrance Journal. 2016;**31**:361-376. DOI: 10.1002/ffj.3325

[33] Madrera RR, Suárez Valles B, Diñeiro García Y, Del Valle Argüelles P, Picinelli Lobo A. Alternative woods for aging distillates - an insight into their phenolic profiles and antioxidant activities. Food Science and Biotechnology. 2010;**19**:1129-1134. DOI: 10.1007/s10068-010-0161-4

[34] Mrvcic J, Posavec S, Kazazic S, Stanzer D, Peša A, Stehlik-Tomas V. Spirit drinks: A source of dietary polyphenols. Croatian Journal of Food Science and Technology. 2012;**4**:102-111

[35] Ivanova-Petropulos V, Balabanova B, Bogeva E, Frentiu T, Ponta M, Senila M, Gulaboski R, Irimie FD. Rapid determination of trace elements in Macedonian grape brandies for their characterization and safety evaluation. Food Analytical Methods. 2017;**10**:459-468. DOI: 10.1007/s12161-016-0604-5

[36] Rodríguez-Solana R, Salgado JM, Domínguez JM, Cortés-Diéguez S. First approach to the analytical characterization of barrel-aged grape marc distillates using phenolic compounds and colour parameters. Food Technolology and Biotechnology. 2014;**52**:391-402. DOI: 10.17113/ftb.52.04.14.3627

[37] Canas S, Caldeira I, Anjos O, Lino J, Soares A, Belchior AP. Physicochemical and sensory evaluation of wine brandies aged using oak and chestnut wood simultaneously in wooden barrels and in stainless steel tanks with staves. International Journal of Food Science and Technology. 2016;**51**:2537-2545. DOI: 10.1111/ijfs.13235

[38] De Rosso M, Cancian D, Panighel A, Dalla Vedova A, Flamini R. Chemical compounds released from five different woods used to make barrels for aging wines and spirits: Volatile compounds and polyphenols. Wood Science and Technology. 2009;**43**:375-385. DOI: 10.1007/s00226-008-0211-8

[39] Schwarz M, Rodríguez MC, Martínez C, Bosquet V, Guillén D, Barroso CG. Antioxidant activity of Brandy de Jerez and other aged distillates, and correlation with their polyphenolic content. Food Chemistry. 2009;**116**:29-33. DOI: 10.1016/j.foodchem.2009.01.096

[40] Rodríguez Dodero MC, Guillén Sánchez DA, Rodríguez MS, Barroso CGJ. Phenolic compounds and furanic derivatives in the characterization and quality control of Brandy de Jerez. Journal of Agricultural and Food Chemistry. 2010;**58**:990-997. DOI: 10.1021/jf902965p

[41] Canas S, Silva V, Belchior PA. Wood related chemical markers of aged wine brandies. Ciência e Técnica Vitivinícola. 2008;**23**:45-52

[42] Madrera RR, Gomis BD, Alonso MJJ. Influence of distillation system, oak wood type, and aging time on composition of cider brandy in phenolic and furanic compounds. Journal of Agricultural and Food Chemistry. 2003;**51**:7969-7973. DOI: 10.1021/jf0347618

[43] Delgado-González MJ, Sánchez-Guillén MM, García-Moreno MV, Rodríguez-Dodero MC, García-Barroso C, Guillén-Sánchez DA. Study of a laboratory-scaled new method for the accelerated continuous ageing of wine spirits by applying ultrasound energy. Ultrasonics Sonochemistry. 2017;**36**:226-235. DOI: 10.1016/j.ultsonch.2016.11.031

[44] Bin Z, Zeng XA, Sun D-W, Shu Juan Y, Yang MF, Ma S. Effect of electric field treatments on brandy aging in oak barrels. Food and Bioprocess Technology. 2013;**6**:1635-1643. DOI: 10.1007/s11947-012-0788-7

[45] Schwarz M, Rodríguez MC, Sánchez M, Guillén DA, Barroso CG. Development of an accelerated aging method for brandy. LWT – Food Science and Technology. 2014;**59**:108-114. DOI: 10.1016/j.lwt.2014.05.060

[46] Caldeira I, Anjos O, Portal V, Belchior AP, Canas S. Sensory and chemical modifications of wine-brandy aged with chestnut and oak wood fragments in comparison to wooden barrels. Analytica Chimica Acta. 2010;**660**:43-52. DOI: 10.1016/j.aca.2009.10.059

[47] Caldeira I, Pereira R, Clímaco MC, Belchior AP, De Sousa RB. Improved method for extraction of aroma compounds in aged brandies and aqueous alcoholic wood extracts using ultrasound. Analytica Chimica Acta. 2004;**513**:125-134. DOI: 10.1016/j.aca.2003.10.011

[48] Caldeira I, De Sousa BR, Belchior PA, Clímaco CM. A sensory and chemical approach to the aroma of wooden aged Lourinhã wine brandy. Ciência e Técnica Vitivinícola. 2008;**23**:97-110

[49] Delgado R, Durán E, Castro R, Natera R, Barroso CG. Development of a stir bar sorptive extraction method coupled to gas chromatography-mass spectrometry for the analysis of volatile compounds in Sherry brandy. Analytica Chimica Acta. 2010;**672**:130-136. DOI: 10.1016/j.aca.2010.05.015

[50] Guerrero ED, Cejudo Bastante MJ, Mejías RC, Marín RN, Barroso CG. Characterization and differentiation of sherry brandies using their aromatic profile. Journal of Agricultural and Food Chemistry. 2011;**59**:2410-2415. DOI: 10.1021/jf104409n

[51] Madrera RR, Suarez Valles B. Determination of volatile compounds în apple pomace by stir bar sorptive extraction and gas chromatography-mass spectrometry (SBSE-GC-MS). Journal of Food Science. 2011;**76**:C1326-C1334. DOI: 10.1111/j.1750-3841.2011.02406.x

[52] Elss S, Preston C, Appel M, Heckel F, Schreier P. Influence of technological processing on apple aroma analysed by high resolution gas chromatography-mass spectrometry and on-line gas chromatography-combustion/pyrolysis-isotope ratio mass spectrometry. Food Chemistry. 2006;**98**:269-276. DOI: 10.1016/j.foodchem.2005.06.011

[53] Wu J, Gao H, Zhao L, Liao X, Chen F, Wang Z, Hu X. Chemical compositional characterization of some apple cultivars. Food Chemistry. 2007;**103**:88-93. DOI: 10.1016/j.foodchem.2006.07.030

[54] Christoph N, Bauer-Christoph C. Flavour of spirit drinks: Raw materials, fermentation, distillation, and ageing. In: Berger RG, editor. Flavours and Fragrances. Berlin, Heidelberg: Springer; 2007. pp. 219-239. DOI: 10.1007/978-3-540-49339-6_10

[55] Schneider R, Charrier F, Moutounet M, Baumes R. Rapid analysis of grape aroma glycoconjugates using Fourier-transform infrared spectrometry and chemometric techniques. Analytica Chimica Acta. 2004;**513**:91-96. DOI: 10.1016/j.aca.2003.11.082

[56] Tietel Z, Porat R, Weiss K, Ulrich D. Identification of aroma-active compounds în fresh and stored 'Mor' mandarins. International Journal of Food Science and Technology. 2011;**46**:2225-2231. DOI: 10.1111/j.1365-2621.2011.02740.x

[57] Villatoro C, Altisent R, Echeverria G, Graell J, Lopez ML, Lara I. Changes în biosynthesis of aroma volatile compounds during on-tree maturation of 'Pink lady (R)' apples. Postharvest Biology and Technology. 2008;**47**:286-295. DOI: 10.1016/j.postharvbio.2007.07.003

[58] Hashimoto A, Kozima K, Sakamoto K. Validity of SPME cryofocusing în analysis of volatile components în orange juice. Food Science and Technology Research. 2006;**12**:295-298. DOI: 10.3136/fstr.12.295

[59] Huber GM, Rupasinghe HPV. Phenolic profiles and antioxidant properties of apple skin extracts. Journal of Food Science. 2009;**74**:C693-C700. DOI: 10.1111/j.1750-3841.2009.01356.x

[60] Kammerer DR, Saleh ZS, Carle R, Stanley RA. Adsorptive recovery of phenolic compounds from apple juice. European Food Research and Technology. 2007;**224**:605-613. DOI: 10.1007/s00217-006-0346-5

[61] Gil MI, Tomas-Barberan F, Hess-Pierce B, Kader AA. Antioxidant capacities, phenolic compounds, carotenoids, and vitamin C contents of nectarine, peach, and plum cultivars from California. Journal of Agricultural and Food Chemistry. 2002;**50**:4976-4982. DOI: 10.1021/jf020136b

[62] Blanco-Gomis D, Alonso JJM, Cabrales IM, Abrodo PA. Characterization of cider apples on the basis of their fatty acid profiles. Journal of Agricultural and Food Chemistry. 2002;**50**:1097-1100. DOI: 10.1021/jf010998x

[63] Fromm M, Bayha S, Carle R, Kammerer D. Comparison of fatty acid profiles and contents of seed oils recovered from dessert and cider apples and further Rosaceous plants. European Food Research and Technology. 2012;**234**:1033-1041. DOI: 10.1007/s00217-012-1709-8

[64] Ozdemir F, Topuz A. Changes în dry matter, oil content and fatty acids composition of avocado during harvesting time and post-harvesting ripening period. Food Chemistry. 2004;**86**:79-83. DOI: 10.1016/j.foodchem.2003.08.012

[65] Cortés S, Salgado JM, Rodríguez N, Domínguez JM. The storage of grape marc: Limiting factor în the quality of the distillate. Food Control. 2010;**21**:1545-1549. DOI: 10.1016/j.foodcont.2010.04.029

[66] Satora P, Tuszynski T. Biodiversity of yeasts during plum Wegierka Zwykla spontaneous fermentation. Food Technology and Biotechnology. 2005;**43**:277-282

[67] Ribeiro LS, Duarte WF, Dias DR, Schwan RF. Fermented sugarcane and pineapple beverage produced using *Saccharomyces cerevisiae* and non-*Saccharomyces* yeast. Journal of the Institute of Brewing. 2015;**121**:262-272. DOI: 10.1002/jib.218

[68] Budić-Leto I, Humar I, Zdunić G, Hribar J, Zlatić E. Volatile compounds in Prošek dessert wines produced from white and red grapes. Czech Journal of Food Sciences. 2015;**33**:354-360. DOI: 10.17221/28/2015-CJFS

[69] Gomes FCO, Silva LC, Vianna CR, Lacerda ICA, Borelli BM, Nunes AC, Franco GR, Mourao MM, Rosa CA. Identification of lactic acid bacteria associated with traditional cachaca fermentations. Brazilian Journal of Microbiology. 2010;**41**:486-492. DOI: 10.1590/S1517-83822010000200031

[70] Geroyiannaki M, Komaitis ME, Stavrakas DE, Polysiou M, Athanasopoulos PE, Spanos M. Evaluation of acetaldehyde and methanol în greek traditional alcoholic beverages from varietal fermented grape pomaces (*Vitis vinifera* L.). Food Control. 2007;**18**:988-995. DOI: 10.1016/j.foodcont.2006.06.005

[71] García-Llobodanin L, Ferrando M, Güell C, López F. Pear distillates: Influence of the raw material used on final quality. European Food Research and Technology. 2008;**228**:75-82. DOI: 10.1007/s00217-008-0908-9

[72] Diban N, Voinea OC, Urtiaga A, Ortiz I. Vacuum membrane distillation of the main pear aroma compound: Experimental study and mass transfer modeling. Journal of Membrane Science. 2009;**326**:64-75. DOI: 10.1016/j.memsci.2008.09.024

[73] Nunes C, Coimbra MA, Saraiva J, Rocha SM. Study of the volatile components of a candied plum and estimation of their contribution to the aroma. Food Chemistry. 2008;**111**:897-905. DOI: 10.1016/j.foodchem.2008.05.003

[74] Tešević V, Nikićević N, Milosavljević S, Bajić D, Vajs V, Vučković I, Vujisić L, Đorđević I, Stanković M, Veličković M. Characterization of volatile compounds of "Drenja", an alcoholic beverage obtained from the fruits of cornelian cherry. Journal of the Serbian Chemical Society. 2009;**74**:117-128. DOI: 10.2298/JSC0902117T

[75] Apostolopoulou AA, Flouros AI, Demertzis PG, Akrida-Demertzi K. Differences în concentration of principal volatile constituents în traditional Greek distillates. Food Control. 2005;**16**:157-164. DOI: 10.1016/j.foodcont.2004.01.005

[76] Wechgama K, Laopaiboon L, Laopaiboon P. Quantitative analysis of main volatile and other compounds în traditional distilled spirits from Thai rice. Biotechnology. 2008;**7**:718-724. DOI: 10.3923/biotech.2008.718.724

[77] Lukić I, Tomas S, Miličević B, Radeka S, Peršurić Đ. Behaviour of volatile compounds during traditional alembic distillation of fermented Muscat Blanc and Muškat Ruža Porečki grape marcs. Journal of the Institute of Brewing. 2011;**117**:440-450. DOI: 10.1002/j.2050-0416.2011.tb00491.x

[78] García-Llobodanin L, Roca J, López JR, Pérez-Correa JR, López F. The lack of reproducibility of different distillation techniques and its impact on pear spirit composition. International Journal of Food Science & Technology. 2011;**46**:1956-1963. DOI: 10.1111/j.1365-2621.2011.02707.x

[79] Claus M, Berglund K. Fruit brandy production by batch column distillation with reflux. Journal of Food Process Engineering. 2005;**28**:53-67. DOI: 10.1111/j.1745-4530.2005.00377.x

[80] Mosedale JR, Puech JL. Wood maturation of distilled beverages. Trends in Food Science & Technology. 1998;**9**:95-101. DOI: 10.1016/S0924-2244(98)00024-7

[81] Savchuk SA, Nuzhnyi VP, Kolesov GM. Factors affecting the accuracy of the determination of diethyl phthalate in vodka, ethanol, and samples of illegal alcoholic products. Journal of Analytical Chemistry. 2006;**61**:1198-1203. DOI: 10.1134/S1061934806120136

[82] Cabrita MJ, Barrocas Dias C, Costa Freitas AM. Phenolic acids, phenolic aldehydes and furanic derivatives în oak chips: American vs. French oaks. South African Journal of Enology and Viticulture. 2011;**32**:204-210. DOI: 10.21548/32-2-1380

[83] Van Jaarsveld FP, Hattingh S, Minnaar P. Rapid induction of ageing character în brandy products – Part II. Influence of type of oak. South African Journal of Enology and Viticulture. 2009;**30**:24-37. DOI: 10.21548/30-1-1420

[84] Van Jaarsveld FP, Hattingh S. Rapid induction of ageing character in brandy products. Ageing and general overview. South African Journal of Enology and Viticulture. 2012;**33**:225-252

[85] Rodríguez-Solana R, Rodríguez-Freigedo S, Salgado JM, Domínguez JM, Cortés-Diéguez S. Optimisation of accelerated ageing of grape marc distillate on a micro-scale process using a box – Benhken design: Influence of oak origin, fragment size and toast level on the composition of the final product. Australian Journal of Grape and Wine Research. 2016;**23**:5-14. DOI: 10.1111/ajgw.12249

[86] Sanz M, Cadahía E, Esteruelas E, Muñoz AM, Fernández De Simón B, Hernández T, Estrella I. Phenolic compounds in cherry (prunus avium) heartwood with a view to their use in cooperage. Journal of Agricultural and Food Chemistry. 2010;**58**:4907-4914

[87] Gortzi O, Metaxa X, Mantanis G, Lalas S. Effect of artificial ageing using different wood chips on the antioxidant activity, resveratrol and catechin concentration, sensory properties and colour of two Greek red wines. Food Chemistry. 2013;**141**:2887-2895. DOI: 10.1016/j.foodchem.2013.05.051

[88] Alañón ME, Castro-Vázquez L, Díaz-Maroto MC, Hermosín-Gutiérrez I, Gordon MH, Pérez-Coello MS. Antioxidant capacity and phenolic composition of different woods used in cooperage. Food Chemistry. 2011;**129**:1584-1590. DOI: 10.1016/j.foodchem.2011.06.013

[89] Fernández de Simón B, Sanz M, Cadahía E, Martínez J, Esteruelas E, Muñoz AM. Polyphenolic compounds as chemical markers of wine ageing in contact with cherry, chestnut, false acacia, ash and oak wood. Food Chemistry. 2014;**143**:66-76. DOI: 10.1016/j.foodchem.2013.07.096

[90] Flamini R, Vedova AD, Cancian D, Panighel A, De Rosso M. GC/MS-positive ion chemical ionization and MS/MS study of volatile benzene compounds în five different woods used în barrel making. Journal of Mass Spectrometry. 2007;**42**:641-646. DOI: 10.1002/jms.1193

[91] Hernández T, Estrella I, Dueñas M, Fernández de Simón B, Cadahía E. Influence of wood origin în the polyphenolic composition of a Spanish red wine aging în bottle, after storage în barrels of Spanish, French and American oak wood. European Food Research and Technology. 2007;**224**:695-705. DOI: 10.1007/s00217-006-0360-7

[92] Zarzynski P. Identification and quantitative analysis of phenolic compounds naturally occuring în wood of selected European and exotic tree species. Forest Research Papers. 2009;**70**:27-39. DOI: 10.2478/v10111-009-0003-4

[93] Wiśniewska P, Dymerski T, Wardencki W, Namieśnik J. Chemical composition analysis and authentication of whisky. Journal of the Science of Food and Agriculture. 2015;**95**(11):2159-2166

[94] Leopold LF, Leopold N, Diehl HA, Socaciu C. Prediction of Total antioxidant capacity of fruit juices using FTIR spectroscopy and PLS regression. Food Analytical Methods. 2012;**5**(3):405-407

[95] Uríčková V, Sádecká J. Determination of geographical origin of alcoholic beverages using ultraviolet, visible and infrared spectroscopy: A review. Spectrochimica Acta Part A: Molecular and Biomolecular Spectroscopy. 2015;**148**:131-137. DOI: 10.1016/j.saa.2015.03.111

[96] Borràs E, Ferré J, Boqué R, Mestres M, Acena L, Busto O. Data fusion methodologies for food and beverage authentication and quality assessment: A review. Analytica Chimica Acta. 2015;**891**:1-14. DOI: 10.1016/j.aca.2015.04.042

[97] Pontes MJC, Santos SRB, Araújo MCU, Almeida LF, Lima RAC, Gaião EN, Souto UTCP. Classification of distilled alcoholic beverages and verification of adulteration by near infrared spectrometry. Food Research International. 2006;**39**:182-189. DOI: 10.1016/j.foodres.2005.07.005

Sample Preparation for Determination of Bioaccessibility of Essential and Toxic Elements in Legumes

Wagna Piler Carvalho dos Santos and
Rita Maria Weste Nano

Additional information is available at the end of the chapter

http://dx.doi.org/10.5772/intechopen.69850

Abstract

The methods used to estimate the bioavailability of elements have different approaches. These tests are based on selective extraction or simulation of the physiology of the gastrointestinal tract. The sample preparation methods require studies about extraction procedures, thermal treatment, and decomposition of organic matter. The method of decomposing organic matter assisted by microwaves introduced adequate results for most chemical elements in pulses. The content of the elements present in the extracts obtained by employing the method physiologically based extraction test (PBET) is lower than those obtained by simple bioaccessibility extraction test (SBET) due to complexing effects of metal ions. The mineral content in the gastric and intestinal stages can vary significantly with the investigated leguminous species and the elements. The thermal processing can affect the concentrations of the elements analyzed in samples from leguminous species. This results from the heat capacity to change the speciation of chemical elements. The change speciation may modify the solubility and mobility of chemical species under the conditions of the gastrointestinal tract, which alters the bioavailability. In this sense, it can be concluded that the domestic cooking process can influence the nutritional and toxicological potential of pigeon pea, cowpea, and mangalo.

Keywords: legumes, bioaccessibility, sample preparation, ICP OES, minerals

1. Introduction

Grain legumes represent an important food group due to the related nutritional and socioeconomic aspects, especially for Brazil, which occupies the position of major producer and consumer of common bean (*Phaseolus vulgaris* L.). Its nutritional importance is due to the presence of important nutrients, such as proteins, vitamins, and minerals. Among these, we highlight the essential minerals present in larger quantities known as macroelements (Ca, K, Mg, P) and trace elements (Co, Cr, Cu, Fe, Mn, and Mo). Knowledge of the total concentration of these elements does not provide sufficient nutritional information for the elucidation of absorption mechanisms as essential nutrients. On the other hand, toxic elements, such as Pb and Cd, may also be present, being considered contaminants. Thus, studies on the bioavailability of nutrients and contaminants are needed to ensure food safety.

Bioavailability and chemical speciation are multidisciplinary areas, which have gained space in the scientific community in recent years. Research into the interactions of various chemical forms, the presence of antinutritional agents, and their absorption into living organisms involves many scientific fields [1–3]. The bioaccessibility and bioavailability of different species of essential or toxic elements are also important because the essentiality depends on the chemical form of the element that is absorbed as well as its toxic potential [1].

The methods employed to determine the bioavailability of a chemical species in the human body are generally very laborious and provide results that are discordant to each other or are not comparable. Therefore, the development and standardization of analytical processes are activities of interest, since they can contribute with advances in the understanding of natural processes related to the environment and nutrition.

The legumes belong to the species *P. vulgaris* L., the botanical family Leguminosae. Intake of legumes (e.g., common beans, lentils, peas, cowpeas) in the Brazilian diet should be encouraged. However, the total amount of a nutrient does not reflect the amount available to the body through absorption. The accessibility of a chemical species to normal metabolic and physiological processes is known as bioavailability. For the assessment of bioaccessibility, the composition of the food should be considered. The legumes have several antinutritional factors that negatively interfere in the bioavailability of elements [2]. The phytic acid (and phytates) is known as a food inhibitor which chelates micronutrient and prevents it to be bioavailable for monogastric animals, including humans, because they lack enzyme phytase in their digestive tract [3].

The bioavailability can be divided into three phases: availability in the intestinal lumen by absorption, adsorption and/or retention in the body, and use by the body. Several factors may influence the bioavailability of minerals, which may be of dietary or physiological origin [4, 5]. The bioavailability of a chemical species can be estimated by means of the percentage of bioaccessibility of this species. Bioaccessibility assays performed using in vitro methods are the focus of this work.

The oral bioaccessibility of a substance can be defined as the fraction soluble in the conditions of the gastrointestinal tract and that is available for absorption [6, 7]. However, some nutrients do not need to be digested to be absorbed and others, even hydrolysates, cannot be absorbed. Iron may be strongly bound in the absorbed chelate structure, with no release of the metal ion to the cells and incorporation by the proteins [8].

Many factors and promoters act on the bioavailability of trace elements, such as chemical form of the mineral in the food, food binders, redox activity in food components, interactions between the minerals, and the individual's physiological state [2, 6]. Therefore, the concept of bioavailability of micronutrients should recognize all important factors, as well as the rates of use of the absorbed nutrient and the rates of exchange and excretion, which can vary considerably, due to (i) intrinsic factors, that is, mechanisms of absorption and metabolic processes and mutual interactions, and (ii) extrinsic, such as solubility, size of molecules, and synergistic or antagonistic chemical effects [6].

Thus, in 1997, at the International Bioavailability Conference, the term bioavailability began to refer to the fraction of any ingested nutrient with the potential to meet physiological demands on target tissues. In 2001, the concept incorporated three aspects: bioconversion, bioefficacy, and bioefficiency [2].

The methods employed to estimate the bioavailability of elements rely on different approaches [9]. In vivo tests make use of guinea pigs (rabbits, rats, pigs, and monkeys). The bioaccessible fraction of the nutrients is determined by the analysis of the animal's nails, hair, and blood after administration of the diet of interest. These tests require specialized professionals and specific infrastructure for their realization, besides having execution times and high costs [10, 11]. In vitro tests may be attractive due to higher analytical speed and lower cost. These tests are based on selective extraction or simulate the physiology of the gastrointestinal tract and can be classified into two categories, static and dynamic, and do not include the microorganisms present in the digestive tract, nor do they consider the adsorption mechanisms that preferentially occur in the duodenal epithelium [10].

The in vitro bioaccessibility of minerals varies significantly, depending on the mineral and the type of the food matrix [3]. On the other hand, it may also change with variations in the sample preparation step. The effects of heat treatment on the bioavailability of some minerals in food matrices were investigated. Different species of legumes consumed in India were studied (*Cicer arietinum, Phaseolus aureus, Phaseolus mungo, Cajanus cajan, Vigna catjang, P. vulgaris*). Most of the works use gastric simulation with simple extraction tests. In addition, few studies have done any kind of comparison of the results using the in vivo method as a reference [11–15].

The possible explanations for this fact are high cost of the in vivo method, high time of analysis, complexity of the tests, and ethical implications [14]. These aspects make it difficult to perform the experiments, making in vitro tests more attractive, which are based on the selective extraction or simulate the physiology of the gastrointestinal tract.

Due to the limitations of the in vivo assays to estimate the bioavailability of metals, since the 1990s it is recommended to replace them by in vitro methods, requiring, for such development and validation. The in vitro approach enables the faster and more accessible generation of information that allows human health risk assessments related to exposure to a specific toxic agent [7].

Many in vitro methods are employed to estimate the bioaccessibility of certain chemical species. A brief comparison of examples of these methods was presented, in which four are framed as static methods and only one as dynamic [10, 11]. The simplest of these is the simple bioaccessibility extraction test (SBET) method proposed by the Consortium for Research on Solubility and Bioavailability (SBRC). Recognized research groups updating in the area are evaluating the measurement of methods for assessing bioaccessibility, such as the Scientific Group on Methods for a Chemical Safety Assessment (SGOMSEC) was established in 1979 [13]. Also more recently the in vitro method recommended by the European Unified Research Group (BARGE) is used to estimate the bioaccessibility of trace elements [14].

The SBET method was developed based on the work initially described by Ruby et al. and simulates the mobilization of the substances in the gastric conditions of the stomach, disregarding the intestinal compartment [6]. Since it only simulates the gastric phase, it generally provides overestimated bioaccessibility results, due to the low pH of the medium and the absence of an intestinal phase. The development, validation, and standardization of these methods are areas that still demand studies [15].

In vitro laboratory tests to predict the bioavailability of metals from a solid matrix that simulate the physicochemical conditions of solutions found in the stomach and in the human duodenum are called bioaccessibility tests and can be known as physiologically based extraction test (PBET) [16]. These tests do not include the microorganisms present in the digestive tract, nor do they consider the adsorption mechanisms that occur preferentially in the duodenal epithelium [10]. The oral bioaccessibility of a substance was then defined as the fraction that is soluble in gastrointestinal tract conditions and is available for absorption [6, 7]. Bioaccessibility values become very useful in the nutritional analysis of foods when it is considered that any soluble nutrient is susceptible to absorption in the human intestine. Leguminosae (chickpeas, lentils, cowpeas, and green peas) are sources of essential elements, particularly K, P, Ca, Cu, Fe, and Zn [17]. Generally, they provide sufficient amounts of Fe, Ca, and P required in a human diet. However, it is necessary to mention the possible presence of inorganic contaminants in plants, which, although generally inferior to foods of animal origin, should also be investigated, since it poses an imminent risk to health maintenance [18]. In recent work, concentrations of Pb in samples of carioca beans varied between 4.6 and 6.2 $\mu g\ g^{-1}$ [19]. These concentrations are above the maximum tolerance limit (LMT) recommended by the Brazilian legislation for legumes in natura or industrialized (0.50 $\mu g\ g^{-1}$) [20]. According to the Codex Alimentarius, in the 2006 review, these levels were set at 0.2 $\mu g\ g^{-1}$ for Pb in legumes in natura and 1.0 $\mu g\ g^{-1}$ for legumes (processed green peas and beans and peas) [21].

2. Methods and materials

2.1. Collection of samples

The choice of samples (grain legumes) was based on being foods consumed in the Northeast, especially species such as cowpea (*Vigna unguiculata* L. Walp), pigeon pea (*C. cajan* L.), and mangalo (*Lablab purpureus* L. Sweet). Samples of cowpea and pigeon pea were obtained in the municipality of Ipirá, a micro-region of Feira de Santana, Bahia, Brazil. Mangalo samples were obtained in the city of Santo Amaro, metropolitan region of Salvador, Bahia, Brazil.

2.2. Reagents and standard solutions

The concentrations of the working solutions of the trace elements Ba, Ca, Cd, Cr, Cu, Fe, K, Mn, Mo, Mg, P, Pb, and Zn (Merck, Germany) were prepared from stock solutions containing 1000 mg L^{-1}. All solutions were prepared with analytical grade reagents and ultrapure water, specific resistivity of 18.2 MΩ cm^{-1} in purification system Milli-Q® (Millipore, Bedford, MA, USA).

As certified reference materials of grain legumes as well as compatible materials for analysis of bioaccessibility of minerals in food matrices are not available on the market, certified reference materials from National Institute of Standards and Technology (NIST, Gaithersburg, Maryland, USA): 1515 apple leaf and 1570a spinach. These materials were used to evaluate the accuracy of the analytical procedure employed for the determination of the total content of the analytes.

The following reagents were used: hydrochloric acid (Carlo Erba, Italy), nitric acid (Merck, Germany), hydrogen peroxide 30% m/v (Synth, São Paulo, Brazil), and glycine (Vetec, Rio de Janeiro, Brazil). Standard buffer solutions pH 7.00 ± 0.05 and pH 4.00 ± 0.02 (Haloquímica, São Paulo) were used for the calibration of the pH meter. All materials used in the collection, storage, and preparation of the samples were previously washed with detergent and decontaminated with nitric acid (10% v/v) for a minimum period of 24 h and rinsed with ultrapure water, with specific resistivity of 18.2 MΩ cm^{-1}.

2.3. Instrumentation

To perform the thermal treatment of the samples, the following equipment was used: drying oven with Fanem forced circulation, model 520, and oven, Panasonic brand, with an output/consumption power of 900 W/1450 W, a frequency of 2450 MHz, and power output/consumption (resistance) of 950 W/1010 W. For cooking in a pressurized system, Teflon™ coated aluminum pan, whose heat source for cooking the food was a Bunsen nozzle, was used.

For the drying of the samples, a lyophilizer Terroni Fauvel LT 1000/8 (São Carlos, São Paulo) was used. The dried samples were ground in an 8000 M ball mill (Spex Sample Prep, USA). The Tecnal (São Paulo) digester block and ETHOS One microwave (Milestone, Italy) were used for the acid decomposition procedure of the samples.

In the bioaccessibility assays, the vials were incubated in a Tecnal incubator, model TE-420, at 37°C and shaking at 100 rpm.

The optical emission spectrometer with inductively coupled argon plasma (ICP OES) with VISTA PRO axial vision (Varian, Mulgrave, Australia) was used to determine the analytes. This instrument is equipped with a solid-state detector with CCD array (charge-coupled device) and operates at wavelengths in the range of 167–785 nm. It has an end-on gas interface, which with the front flow countercurrent gas protects the pre-optical region from overheating and removes the colder zone from the plasma. The spectral lines were selected considering the intensities of the emission signals of the analytes and the background signal, the standard deviation of the measurements, the adequate sensitivity for the determination of the elements present in high and low concentrations in the matrices, as well as the profile of the spectra and the possibility of interference.

2.4. Preparation of the samples

Pretreatment of samples consisted of selection, washing, freeze-drying for 48 h, milling for 2 min, sieving (nylon mesh, <300 μm), packaging, and storage. The green, fresh, and moist grains of the legumes were washed, selected, drained, packed in polypropylene bottles, and preserved under freezing at −30°C.

The bioaccessibility of the minerals in the legumes was evaluated in the cooking process, for which the samples were heat treated in three types of heating and two time levels: (a) under pressure at 15 psi (3 and 6 min), (b) in the oven (20 and 40 min), and (c) in a microwave oven (6 and 12 min). During cooking, the conditions were determined: sample mass/volume ratio (100 g sample/300 mL water), oven temperature 2000°C, and 100% microwave power. For oven and microwave oven processes, the cooking of the samples was performed in beakers of 1000 mL. After the cooking time set for each experiment, the grains were drained in a plastic sieve. This stage simulates the domestic procedure that is usually performed in the preparation of these foods. For comparison of the results, samples that were not heat treated were also analyzed. For each species three sub-samples were produced. After cooking the samples were dried in an oven with air circulation at 60°C and ground in a ball mill. The dried and ground samples were stored in decontaminated plastic bottles and kept at refrigeration temperature (<10°C).

2.5. Sample acid decomposition procedure

For the digestion block acid decomposition procedure, 500 mg of the previously dried and ground samples was weighed directly into digestion tubes, and 5.0 mL of 65% w/w nitric acid was added. The digestion was started with a gradual increase in temperature, starting at 50°C, rising to 100°C, and ending at 150°C, maintaining this temperature for 30 min. At the time the temperature reached 150°C, hydrogen peroxide (H_2O_2) was slowly added in 1 mL portions. At the end of the 15 min, 10 mL peroxide was added to each tube. To promote the condensation and reflux of gases and vapors generated in the digestion, minimizing contamination and loss of the volatile chemical elements, cold fingers were adapted to the cold digestion flasks [22]. At the end of the digestion, due to the presence of particulate, the solution obtained was filtered on medium filtration filter paper into 25 mL volumetric flasks. The volume of the flask was filled with ultrapure water, with resistivity of 18.2 MΩ cm^{-1}.

A microwave-assisted decomposition procedure was also employed. In the procedure, 9.0 mL of nitric acid, 4.0 mol L^{-1}, 1.0 mL of hydrogen peroxide 30% v/v, and 500 mg of the sample were used. The heating program consisted of four steps: (1) ramp of 6 min and temperature of 90°C, (2) 5 min at 90°C, (3) ramp of 10 min and temperature of 190°C, and (4) 10 min at 190°C. The volume of the digested mixture was adjusted to 15.0 mL with ultrapure water. The solutions were stored in polypropylene bottles at refrigeration temperature.

2.6. Procedure for bioaccessibility testing

2.6.1. Simple bioaccessibility extraction test (SBET)

The bioaccessibility assays were conducted by the SBET method with mangalo pulse samples. For this assay, a 0.4 mol L-1 glycine solution in acid medium (HCl, pH = 1.5) was used to simulate gastric digestion. To this, 0.25 g of sample was incubated with the prepared solution; the temperature was adjusted to 37°C and a rotation speed to 100 rpm for 1 h. Subsequently, the mixture was vacuum filtered on filter membranes of cellulose acetate having porosity of 0.45 μm (Millipore). The extracts were stored in plastic vials under refrigeration temperature for a maximum time of 24 h.

The use of these membranes increases the time of analysis and may increase the risk of contamination. In this way, the influence of the separation of the solid phase (legume sample) from the solution (bioaccessible fraction), obtained after sample incubation stage, was evaluated. To this end, the membrane filtration step was replaced by centrifugation at 3000 rpm for 15 min. Of the sample-solution mixture, the bioaccessibility of Cu, Mn, and Zn was determined. Samples from the three species were used for this evaluation.

2.6.2. Physiologically based extraction test (PBET)

To estimate the analytes' bioavailability, the PBET method was used, which involves the simulation of gastric digestion conditions, followed by the simulation of intestinal digestion conditions [23, 24]. For this, the sample was incubated at 37°C for 1 h with pepsin at pH 2.5 (simulated gastric digestion) and then at pH 7.0 with pancreatin and bile extract (simulating intestinal digestion). Each batch consisted on average of four samples selected randomly, in triplicate of gastric and intestinal digestion, plus three gastric digestion blank samples and three intestinal digestion blank samples, totaling 30 beakers. Each beaker (except the blank samples) contained 0.300 g of the sample and 30.0 mL of gastric solution prepared on the day. Gastric solution was prepared with 1.25 g pepsin, 500 mg malate, 500 mg of citrate, 500 uL of acetic acid, and 420 uL of lactic acid diluted in 1.0 L of ultrapure water, and the pH was adjusted to 2.5 with hydrochloric acid solution. The mixture was incubated at 37°C under orbital shaking at 100 rpm for 1 h. The pH of the mixture was adjusted to 7.0 with $NaHCO_3$ solution. To the intestinal phase, 15 mg of pancreatin and 52.5 mg of bile salts were added, and this mixture was incubated for 4 h under the same conditions of the gastric phase.

2.7. Determination of analytes

The concentration of each analyte in the digests after acid decomposition of the samples and in the extracts of the bioaccessibility assays was obtained using inductively coupled plasma optical emission spectrometry (ICP OES). The optical system of the ICP OES was calibrated with a multielement reference solution, and torch alignment was performed with a 5.0 mg Mn/L solution. The spectral lines were selected considering the intensities of the emission signals of the analytes and the background signal, the standard deviation of the measurements, the adequate sensitivity for the determination of the elements present in high and low concentrations in the matrices, as well as the profile of the spectra and the possibility of interference.

3. Results and discussions

3.1. Determination of mineral contents

Initially, the accuracy of the procedure used to determine the total analyte content was checked using CRM 1515 and 1570a. The average extraction efficiency and RSD of K, P, Ba, Cu, Mn, and Zn were, respectively, 93% and less than 5%. The lowest extraction efficiency was due to Fe in CRM 1515 (83%).

The traces of Ba, Cu, Fe, Mn, and Zn were determined in the cowpea, pigeon pea, and mangalo samples after acidic decomposition. The samples did not undergo any heat treatment except drying the grains in a greenhouse with 60°C air circulation and milling. The concentrations in mg kg^{-1} and interval at the 95% confidence level of the metals were Ba 8.9 ± 0.1, Cu 8.90 ± 0.1, Fe 37.1 ± 0.4, Mn 16.7 ± 0.2, and Zn 30.6 ± 0.4 for cowpea; Ba 0.07 ± 0.01, Cu 3.94 ± 0.01, Fe 50.2 ± 0.06, Mn 23.6 ± 0.04, and Zn 47.3 ± 0.7 for Pigeon pea; and Ba 1.50 ± 0.01, Cu 11.4 ± 0.01, Fe 82.9 ± 0.1, Mn 28.9 ± 0.01, and Zn 43.5 ± 0.2 for mangalo.

The contents of the major elements Ca, K, Mg, and P were determined in the samples. The concentrations in mg kg^{-1} and interval at the 95% confidence level of the metals were Ca 817 ± 8, K 137,914 ± 792, Mg 1155 ± 9, and P 2086 ± 27 for cowpea; Ca 829 ± 3, K 137,914 ± 68, Mg 1785 ± 2, and P 5149 ± 4 for pigeon pea; and Ca 356 ± 1, K 183,923 ± 150, Mg 1975 ± 1e, and P 5939 ± 7 for mangalo.

3.2. Step of separation of the in vitro method

The ratio between the content of the analyte present in the sample and the filtrate corresponds to the bioaccessibility of the minerals. This ratio is then multiplied by a factor of 100, and the result is expressed as a percentage of bioaccessibility, as shown in Eq. (1):

$$\%B = \frac{Y}{Z} \times 100 \tag{1}$$

where Y is the content of the element in the bioaccessible fraction and Z is the total content of the element [12].

The bioaccessibility of a chemical species can be defined as the fraction soluble in the conditions of the gastrointestinal tract and that is available for absorption. In in vitro methods, as in the SBET method, 0.45 μm membrane filters are used. The filtration stage aims to separate the fraction dissolved in the gastric simulation of the solid phase.

The two separation procedures were compared: (a) vacuum filtration in cellulose acetate membrane of 0.45 μm porosity and (b) centrifugation at 3000 rpm for 15 min. Analytes were determined using ICP OES.

The bioaccessibility results of the solid phase separation tests using membrane filtration and centrifugation of three bean species showed the following intervals in membrane filtration: Cu 93–111%, Ba 0–41%, Fe 40–58%, Mn 97–116%, Zn 105–106%, Ca 97–121%, K 95–97%, Mg

101–103%, and P 62–75%. The results for solid phase separation using centrifugation were Cu 94–113%, Ba 0–41%, Fe 39–52%, Mn 107–111%, Zn 105–109%, Ca 98–107%, K 98–100%, Mg 102–108%, and P 64–76%.

The results were analyzed by applying a paired *t*-test at a 95% confidence level. The comparison between the variances of the methods was analyzed using an *F* test. No significant difference ($p > 0.05$) was observed between the means ($n = 3$) of the percentages of bioaccessibility of Cu and Zn, obtained with the procedure employing centrifugation, when compared to the filtration procedure.

3.3. Bioaccessibility: gastric and intestinal phases

In vitro methods that simulate only the gastric phase give limited information about the potential bioavailability of the nutrient or toxic species. The PBET in vitro method simulates gastric conditions including the use of enzymes from this compartment as well as simulates the intestinal compartment promoting alkaline pH change and biliary enzyme addition. For the determination of the bioaccessibility of the minerals, the PBET method was used for the cowpea, pigeon pea, and mangalo samples. The results presented in **Table 1** refer to the percentages of bioaccessibility of Ca, Cu, Fe, K, Mg, and Zn in samples previously dried in an oven with 60°C air circulation and ground.

It is observed that potassium and iron presented the lowest percentage of bioaccessibility considering the three legumes species investigated. It is observed that iron presented the lowest bioaccessibility in the pigeon pea sample. Also, this legume presented the lowest bioaccessibility of potassium when compared to mangalo and cowpea. Mangalo had the lowest bioaccessibility for calcium (39.0%) and the opposite behavior for magnesium (92.3%). Cowpea also presented high bioaccessibility of Mg (97.0). The percentages of bioaccessibility for the Cu and Zn elements varied in a relatively narrow range, comparing with the other elements and the three legumes.

Concentrations of Cu, Fe and Zn, Ca, K, and Mg were also determined in the extracts of the gastric phase and intestinal phase for comparison purposes. The percentage of bioacessible Fe in the intestinal phase varied, respectively: cowpea, 0% at 27 ± 4%; mangalo 0% at 1.8 ± 0.7%;

Elements	% of bioaccessibility		
	Cowpea	Pigeon pea	Mangalo
Ca	79.4	70.3	39.7
K	17.0	7.1	13.0
Mg	97.0	66.1	92.3
Cu	57.8	49.9	62.3
Fe	24.8	3.1	39.9
Zn	43.7	39.3	52.4

Table 1. Mean bioaccessibility of trace elements and higher in legume samples using the PBET method.

and pigeon pea, 14 ± 2% at 83 ± 3%. And the percentages for Cu, 0% at 119 ± 14% (cowpea), 0% at 35 ± 9% (mangalo), and 0% at 76 ± 3% (pigeon pea).

However, the bioavailability of Zn was 52 ± 7% and 36 ± 1% (mangalo) and 74 ± 3% and 70 ± 1% (pigeon pea) for the gastric and intestinal phases, respectively. The bioaccessibility of Mg, for all species, was higher in the intestinal phase.

Among the trace elements, it was observed that copper presented greater bioavailability (100%) in the intestinal phase, accompanied by iron and zinc, which also possessed higher bioavailability in the intestinal phase. This increased bioavailability may be explained according to the level of protein and carbohydrate aggregation in the in the gastric and intestinal phases. Protein degradation begins in the stomach, but it is only complete with the enzymes present in the intestine, whereas carbohydrate degradation begins in the mouth with the salivary amylase, is interrupted in the stomach, and continues again in the intestine, in the form of amino acids and glucose. If there was any fraction of "bound" metal in a protein or carbohydrate, it would now be "released" due to the breakdown of protein or carbohydrate.

For Ca, the gastric and intestinal phases were 58 ± 4% and 98 ± 4% (cowpea) and 89 ± 4% and 56 ± 3% (pigeon pea). For the mangalo sample, the bioaccessibility of Ca did not vary between phases (42 ± 2). The percentages of K in the gaseous phase were higher for pigeon pea (11.9 ± 0.4) and mangrove (8.2 ± 0.2).

Considering the extractions of the major elements, it was observed that it is in the intestinal phase that most of the absorption takes place. These elements, for the most part, are associated with carbohydrates and proteins, so the higher the level of breakage of these molecules, the more macroelements will be bioavailable.

3.4. Comparison of SBET and PBET methods

We compared the results between the SBET method and the PBET method, and it was observed that, among the trace elements, the higher extraction occurs in the SBET because the gastric medium is more acidic and it is easier to have these ions in solution. In this method, the gastric compartment was simulated with an extractor liquid containing only 0.4 mol L^{-1} glycine solution at pH 1.5 adjusted with concentrated hydrochloric acid. In the PBET method, pepsin, malate and citrate are used.

The chemical changes necessary for the digestive process are achieved with the aid of digestive tract enzymes. These enzymes catalyze the hydrolysis of native proteins into amino acids, from starches to monosaccharides and from fats to glycerol and fatty acids. During these digestive reactions, minerals and fat-soluble vitamins in food may become more available for metabolic functions [7].

Therefore, it is expected that the trace element contents present in the extracts obtained using the PBET method will be lower than those obtained by the SBET due to the complexing effects of the metal ions. Among copper, iron, and zinc, it was observed that, in iron, the highest percentage of extraction was mostly (100%) in the SBET, in the copper 50% of the samples extracted more in the SBET, and 50% in the PBET and the zinc was the only one that most of the samples had higher extraction percentages in PBET.

As expected, there was a higher percentage of extraction in the PBET when compared to the trace elements; however, the potassium had the highest extraction percentages (100%) in the SBET. Calcium showed high percentages also in the SBET, the opposite behavior observed for the manganese that presented higher percentages of extraction by the PBET method.

3.5. Cooking effect

Initially, comparing the analyte contents of the cooked samples with respect to the uncooked samples, significant differences were observed at the 95% confidence level in the concentrations of nickel, molybdenum, and barium. The levels of Ni, Se, Mo, Sn, and Ba differed significantly ($p < 0.05$) for analyte concentrations for the three legume species investigated.

With the analysis of the results, it is possible to observe that there is variation of the concentrations according to the fact that the samples are not processed thermally and according to the duration of the cooking. This fact was observed because the heat treatment can influence in the form, that is, in the speciation with which the chemical species presents itself in the food. From this influence the thermal processing can alter the mobility and the solubility of the elements in the conditions of the gastrointestinal tract, thus interfering in the bioaccessibility.

The alteration of speciation in the cooking process occurs because this process is capable of causing separation effects, fractionation of minerals, destruction of inhibitors, formation of complexes with metal ions, denaturation of enzymes that degrade inhibitors, or generation of compounds insoluble by oxidation and precipitation [25]. This ability to change the speciation of the elements also warrants observations on changes in analyte concentrations as the duration of the heating has been greater or less. With the rise of the heating period, the greater the power supply, therefore the more prone is the speciation change. In the majority of sample, it was verified that the longer the thermal processing time, the higher the analyte concentrations in the legume samples.

To evaluate the effect of cooking on the bioaccessibility of minerals, tests were performed using the SBET method. The statistical evaluation of the effect of the thermal treatment on the bioaccessibility of the minerals was performed by paired *t*-test, in which the following results were obtained: at the 95% confidence level.

Using oven cooking, the cooking time significantly interferes, at 95% confidence, in the bioaccessible percentage of Cu. However, the bioavailability of Zn and Fe is not significantly affected by changing the cooking time in the oven. When using the microwave oven for cooking, the bioaccessible percentage of Fe, Cu, and Zn does not differ significantly, at 95% confidence, with different cooking times being achieved. On the other hand, in the cooking under pressure, the bioavailability of Cu differs significantly, at 95% confidence, using different cooking times, since the bioavailability of Zn and Fe does not differ significantly for different cooking times.

Comparing the bioaccessibility percentages obtained in cooked and raw samples, there was a significant difference for Ba, Cu, Fe, K, P, and Zn. Higher percentages of bioaccessibility were obtained for Fe, K, and P in the extract of the unprocessed samples. However, opposite behavior was observed for Cu, i.e., thermal processing favored the availability of this element. However, the behavior of Ba and Zn differed from the others, since for the samples processed in microwave oven and under pressure, the bioaccessibility was lower for Ba and higher for Zn, when compared to the unprocessed samples. The species variable was also significant for these elements.

4. Conclusions

The study carried out between the SBET extraction procedure showed that centrifugation presented comparable results to the conventional method, which employs filtration. Thus, the proposed method for centrifugation is a safe, faster, and lower cost alternative for a phase separation of the SBET method.

Comparison of the traceability and major percentage bioaccessibility percentages using the SBET and PBET methods in the samples of the three legume species confirmed that smaller amounts of minerals are extracted when using the method that simulates gastrointestinal conditions when compared to the method that simulates only the condition of the stomach compartment. This leads in fact, in some cases, to overestimations of bioaccessibility. However, this behavior cannot be generalized, since it can vary from element to element and between food matrices.

The results suggest that the thermal processes investigated can influence the bioaccessibility of the macroelements (Ca, Mg, K, and P) and trace elements (Ba, Cu, Fe, and Zn) in the bioaccessibility of the minerals, i.e., thermal processing may result in increase or decrease in the bioaccessibility of the element.

Acknowledgements

The authors are grateful to the Fundação de Amparo a Pesquisa do Estado da Bahia (FAPESB), Conselho Nacional de Desenvolvimento Científico e Tecnológico (CNPq), and the Instituto Federal da bahia, for the grants and for their financial support.

Author details

Wagna Piler Carvalho dos Santos* and Rita Maria Weste Nano

*Address all correspondence to: wagna.ifba@gmail.com

Instituto Federal da Bahia (IFBA), Salvador, Brazil

References

[1] Moreda-Piñeiroa J, Moreda-Piñeirob A, Romarís-Hortasb V, Moscoso-Péreza C, López-Mahíaa P, Muniategui-Lorenzoa S, Bermejo-Barrerab P, Prada-Rodrígueza D. In-vivo and in-vitro testing to assess the bioaccessibility and the bioavailability of arsenic, selenium and mercury species in food samples. Trends in Analytical Chemistry. 2011;**30**(2):324-345. DOI: 10.1016/j.trac.2010.09.008

[2] Cozzolino SMF. Biodisponibilidade de nutrientes. 3rd ed. São Paulo: Manole; 2009. DOI: 10.1590/S1516-93322005000100017

[3] Gupta RK, Gangoliya SS, Singh NK. Reduction of phytic acid and enhancement of bioavailable micronutrients in food grains. Journal of Food Science and Technology. 2015;**52**(2):676-684. DOI: 10.1007/s13197-013-0978-y

[4] House WA. Trace element bioavailability as exemplified by iron and zinc. Field Crops Research. 1999;**60**:115-141. DOI: 10.1016/S0378-4290(98)00136-1

[5] Peakall D, Burges J. Methodologies for assessing to metals: Speciation, bioavailability of metals, and ecological host factors. Ecotoxicology and Environmental Safety. 2003;**56**:110-121. DOI: 10.1016/S0147-6513(03)00055-1

[6] Ruby MV, Davis A, Kempton JH, Drexler JW, Bergstrom PD. Lead bioavailability: Dissolution kinetics under simulated gastric conditions. Environmental Science & Technology. 1992;**26**:1242-1248. DOI: 10.1021/es50002a614

[7] Ruby MV, Schoof R, Brattin W, Goldade M, Post G, Harnois M, Mosby DE, Casteel SW, Berti W, Carpenter M, Edwards D, Cragin D, Chappell DW. Advances in evaluating the oral bioavailability of inorganics in soil for use in human health risk assessment. Environmental Science & Technology. 1999;**33**:3697-3705. DOI: 10.1021/es990479z

[8] Fennema OR, Parkin K, Damodaran S. Food Chemistry. 4th ed. Boca Raton: CRC Press; 2008. p. 1160. ISBN: 9780849392726

[9] Nayak B, Madhavan Nair K. In vitro bioavailability of iron from wheat flour fortified with ascorbic acid, EDTA and sodium hexametaphosphate, with or without iron. Food Chemistry. 2003;**80**:545-550. DOI: 10.1016/S0308-8146(02)00341-2

[10] Bosso ST, Enzweiler J. Essays for determining lead (Bio)availability in contaminated soils: Review. Química Nova. 2008;**31**:394-400. DOI: 10.1590/S0100-40422008000200036

[11] Oomen AG, Hack A, Minekus M, Zeijdner E, Cornelis SC, Schoeters G, Verstraete W, Van De Wiele T, Wragg J, Rompelberg CJM, Sips A, Van Wijnen JH. Comparison of five in-vitro digestion model to study the bioaccessibility of soil contaminants. Environmental Science & Technology. 2002;**36**:3326. DOI: 10.1021/es010204v

[12] Hemalatha S, Platel K, Srinivasan K. Influence of heat processing on the bioaccessibility of zinc and iron from cereals and pulses consumed in India. Journal of Trace Elements in Medicine and Biology. 2007;**21**:1-7. DOI: 10.1016/j.jtemb.2006.10.002

[13] Goldstein BD. Methodologies for assessing exposures to metals: Speciation, bioaccessibility, and bioavailability in the environment, food, and feed. Ecotoxicology and Environmental Safety. 2003;**56**(1):1-2. DOI: 10.1016/S0147-6513(03)00045-9

[14] Tokalioğlu S, Clough R, Foulkes M, Worsfold P. Bioaccessibility of Cr, Cu, Fe, Mg, Mn, Mo, Se and Zn from nutritional supplements by the unified BARGE method. Food Chemistry. 2014;**150**:321-327. DOI: 10.1016/j.foodchem.2013.10.151

[15] Intawongse M, Dean JR. In-vitro testing for assessing oral bioaccessibility of trace metals in soil and food samples. Trends in Analytical Chemistry. 2006;**25**(9):876-886. DOI: 10.1016/j.trac.2006.03.010

[16] Ruby MV, Davis A, Schoof R, Eberle S, Sellstone CM. Estimation of lead and arsenic bioavailability using a physiologically based extraction test. Environmental Science & Technology. 1996;**30**:422-430. DOI: 10.1021/es950057z

[17] Iqbal A, Khalil IA, Ateeq N, Khan MS. Nutritional quality of important food legumes. Food Chemistry. 2006;**97**:331-335. DOI: 10.1016/j.foodchem.2005.05.011

[18] Kahlon TS, Smith GE, Shao Q. In vitro binding of acids by kidney bean (*Phaseolus vulgaris*), black gram (*Vigna mungo*), Bengal gram (*Cicer arietinum*) and moth bean (*Phaseolus aconitifolins*). Food Chemistry. 2005;**90**:241-246. DOI: 10.1016/j.foodchem.2004.03.046

[19] Santos WPC, Gramacho DR, Teixeira AP, Costa ACS, Korn MGA. Use of Doehlert design for optimizing the digestion of beans for multi-element determination by inductively coupled plasma optical emission spectrometry. Journal of the Brazilian Chemical Society (Impresso). 2008;**19**:1-10. DOI: 10.1590/S0103-50532008000100002

[20] Anvisa. Agência Nacional de Vigilância Sanitária. Sistema de Legislação em Vigilância Sanitária (VISALEGIS). Resolução RDC No. 42/2013. Ministério da Saúde. Agência. Available from: http://www.agricultura.gov.br/arq_editor/file/CRC/Resolu%C3%A7%C3%A3o%2042%20-%20ANVISA.pdf [Accessed: January 4, 2015]

[21] Codex Alimentarius. Codex General Standard for Contaminants and Toxins in Foods; CODEX STAN 193-1995, Rev.2-2006. Available from: http://www.codexalimentarius.net/download/standards/17/CXS_193e.pdf [Accessed: October 3, 2014]

[22] Nano RMW, Bruns RE, Ferreira SLC, Baccan N, Sousa RA, Cadore S. Statistical mixture design development of digestion methods for Oyster tissue using inductively coupled plasma optical emission spectrometry for the determination of metallic ions. Talanta. 2009;**80**:559-564. DOI: 10.1016/j.talanta.2009.07.025

[23] Navarro P, Arana G, Etxebarria N, Dean JR. Evaluation of the physiologically based extraction test as an indicator of metal toxicity in mussel tissue. Analytica Chimica Acta. 2008;**622**:126-132. DOI: 10.1016/j.aca.2008.05.052

[24] Turner A, Radford A. Bioaccessibility of trace metals in boat paint particles. Ecotoxicology and Environmental Safety. 2010;**73**:817-824. http://dx.doi.org/10.1016/j.ecoenv.2010.02.022

[25] Watzke HJ. Impact of processing on bfioavailability examples of minerals in food. Review. Trends in Food Science & Technology. 1998;**9**:320-327. DOI: 10.1016/S0924-2244(98)00060-0

Chapter 8

Sample Preparation Methods for Pesticide Analysis in Food Commodities, Biological and Environment Matrices

Renata Raina-Fulton and Zhen Xie

Additional information is available at the end of the chapter

http://dx.doi.org/10.5772/intechopen.69791

Abstract

This chapter focuses on sample preparation procedures for pesticide analysis of food commodities, biological and environmental matrices. This will include pesticides with a broad range of polarity including those that are more amenable to gas chromatography-mass spectrometry (organochlorines, organophosphorus pesticides, and pyrethroids) and those commonly analyzed by liquid chromatography-mass spectrometry (carbamates, azole, and strobilurin fungicides, and phenylureas as well as organophosphorus pesticides). QuEChERS (quick, easy, cheap, effective, rugged, and safe) methods or QuEChERS methods with modifications to allow wetting of the dry sample matrix, buffering, changing extraction solvent from acetonitrile to ethyl acetate are examined. Subsequent cleanup using dispersive solid phase extraction or cartridge format solid phase extraction has also been completed to reduce matrix effects. Other solid matrices are frequently extracted with pressurized liquid extraction, microwave assisted extraction, or ultrasonic extraction combined with or followed by dispersive solid phase extraction or solid phase extraction. Particularly for chromatography-mass spectrometry, careful consideration of matrix effects needs to be made when considering the design of the sample preparation procedures. Selection of extraction solvent needs to consider both polarity of target analytes (and their solubility in selected solvents) as well as co-extracted matrix components.

Keywords: QuEChERS (quick, easy, cheap, effective, rugged, and safe) methods, solid phase extraction, gas chromatography-mass spectrometry (GC-MS), gas chromatography-tandem mass spectrometry (GC-MS/MS), liquid chromatography-tandem mass spectrometry (LC-MS/MS), azole fungicides, carbamates, organochlorines, organophosphorus pesticides (OPs), phenylureas, pyrethroids, strobilurin fungicides, metabolites, degradation products

1. Introduction

Gas chromatography-mass spectrometry (GC-MS), gas chromatography-tandem mass spectrometry (GC-MS/MS), and liquid chromatography-tandem mass spectrometry (LC-MS/MS) methods are used to analyze for azoles, carbamates, organophosphorus pesticides, pyrethroids, phenylureas, strobilurin fungicides, and other pesticides in a diverse range of sample matrices including food commodities, biological and environmental matrices. The chromatography-mass spectrometry choices for the analysis of these pesticides and others have been recently reviewed [1, 2]. Briefly, organochlorines (OCs), organophosphorus pesticides (OPs), and pyrethroids are frequently analyzed with GC-MS or GC-MS/MS methods. Analysis of azole fungicides, carbamates, neonicotinoids, phenylureas, and strobilurin fungicides is more often analyzed by LC-MS/MS methods. Use of liquid chromatography-electrospray ionization-mass spectrometry (LC-ESI$^+$-MS/MS) for analysis of OPs has also increased over the last 10 years [1]. This chapter will discuss selection choices for extraction and cleanup of sample extracts or preconcentration of target analytes prior to chemical analysis (chromatography-mass spectrometry methods) to minimize matrix enhancement or suppression observed in MS detection. The options for preconcentration or cleanup of sample extracts also depend upon whether the sample is a liquid or solid matrix, fat content, and water content. Modified QuEChERS and microwave and pressurized solvent extraction remain the most widely used extraction procedures with inclusion or subsequent cleanup using dispersive solid phase extraction (dSPE) or solid phase extraction (SPE) methods and will be the focus of discussions in this chapter.

2. Modified QuEChERS procedures and dispersive solid phase extraction

QuEChERS (quick, easy, cheap, effective, rugged, and safe) methods without buffer or with acetate or citrate buffer or other modified QuEChERS methods remain one of the most popular approaches to sample extraction and cleanup of food commodities (**Table 1**). This approach has also been applied to other solid sample matrices including bee products and soil as shown in **Table 1** [3–28]. **Figure 1** shows a comparison of the typical parameters used in various modified QuEChERS methods. Phase separation and partitioning of target analytes into the organic phase is generally achieved with addition of anhydrous $MgSO_4$ (subsequently noted as $MgSO_4$) and NaCl. Addition of NaCl improves the removal of acetonitrile from the aqueous phase and partitioning of polar analytes into acetonitrile [29]. The salt-out extraction is followed by cleanup of the extract with dispersive solid phase extraction (dSPE). Common dSPE sorbents include C18 or C8 for removal of lipids; florisil for removal of polar and low-fat co-extracts; graphitized carbon black (GCB) for removal of pigments and some fatty acids; primary secondary amine (PSA) for efficient removal of saccharides and organic acid as it is a weak anion exchanger; and Z-Sep (ZrO_2 bonded to silica) or Z-Sep+ (ZrO_2 and C18 both bonded to silica) for removal of lipids [15]. PSA has been reported to remove butanoic acid, decanoic acid, heptanoic acid, hexanoic acid, linoleic acid, and phytosterol (stigmasterol), while not effectively removing alkaloids (caffeine and theobromine) and γ-tocopherol [30]. The use of GCB with PSA, C18, and anhydrous $MgSO_4$ was found to improve recoveries for OPs and carbamates in egg matrix as compared to when GCB was not used [31].

QuEChERS (solvent, salts)	dSPE (solvent)	Sample matrix	Pesticides [recoveries]	Analysis method	Reference
ACN salt-out	PSA:C18:GCB (1:1:1) 50 or 125 mg (ACN followed by ACN/ toluene 3:1)	Pollen and single bumble pees	Neonicotinoids (acetamiprid, clothianidin, imidacloprid, thiacloprid, thiamethoxam) [81–87%, pollen; 88–96% bumble bee]; azoles (epoxiconazole, flusilazole, metconazole, tebuconazole, triticonazole) [81–102% pollen; 75–90% bumble bee]; strobilurin fungicides (fluoxastrobin, pyraclostrobin, trifloxystrobin) [71–87%, pollen; 74–82% bumble bee]; others (boscalid, carbendazim, carboxin, prochloraz, spiroxamine) [66–88%, pollen; 63–90% bumble bee].	LC-ESI$^+$-MS/MS	[3]
7.5 mL H_2O, 10 mL ACN, 6 g $MgSO_4$, 1 g NaCl	15 mg C18, 50 mg PSA, 50 mg $MgSO_4$ per mL of ACN extract	Honey bees	Azoles (imazalil, prochloraz, tebuconazole, thiabendazole) [77–96%]; Carbamates (carbendazim, carbofuran, methiocarb) [70–95%]; neonicotinoids (acetamiprid, imidacloprid, thiamethoxam) [80–92%]; OPs (azinphos ethyl, azinphos methyl, chlorfenvinphos, chlorpyrifos, coumaphos, diazinon, diclofenthion, dimethoate, ethion, fenitrothion, fenthion, malathion, omethoate, parathion-ethyl, parathion-methyl, triclofos-methyl) [70–95%]; phenylureas (diuron, isoproturon) [82–86%]; pyrethroids (flumethrin, fluvalinate) [84–93%]; triazines (atrazine, simazine, terbumeton, terbuthylazine) [80–91%]; Degradation products (atrazine-desethyl, atrazine-desisopropyl, carbofuran-3-hydroxyl, fenoxon-sulphone [70–75%],	LC-ESI$^+$-MS/MS	[4]

QuEChERS (solvent, salts)	dSPE (solvent)	Sample matrix	Pesticides [recoveries]	Analysis method	Reference
			fenoxon-sulfoxide, fenthion-sulfone, fenthion-sulfoxide [75–80%], terbumeton-desethyl, terbuthylazine-2-hydroxyl, terbuthylazine-desethyl [75–82%]) [80–94%].		
10 mL ACN, 3 mL hexane before salt addition		Honey bees Wetted (10 mL H_2O)	OPs (coumaphos, diazinon, dimethoate, heptenophos, methidathion, omethoate, oxydemeton-methyl, profenophos, pyrazophos, temephos) [70–93%]	LC-ESI$^+$-MS/MS	[5]
ACN, 4 g $MgSO_4$, 1 g NaCl, 1 g Na_3citrate dihydrate, 0.5 g Na_2Hcitrate sesquihydrate	PSA (25 mg), 150 mg $MgSO_4$ per mL extract	Pollen	Azole fungicides (bitertanol, bromuconazole, difenoconazole, diniconazole, epoxiconazole, fenbuconazole, flusilazole, flutriafol, hexaconazole, paclobutrazole, penconazole, prochloraz [70%], propiconazole, tetraconazole, etc.) [88–94%]; N-methylcarbamates (carbaryl, formetanate, methomyl, oxamyl, pirimicarb, propoxur) [93–96%]; Neonicotinoids (acetamiprid, clothianidin, imidacloprid, nitenpyram, thiacloprid, thiamethoxam) [average 96 and 107%]; OPs (azinphos methyl, demeton-s-methyl sulfone, diazinon, dicrotophos, dimethoate, ethio, ethoprophos, fenamiphos, fenthion degrades, malaoxon, methamidophos, phenofos, trichlorfon, etc.) [>70%]; Strobilurin fungicides (azoxystrobin, kresoxim-methyl, pyraclostrobin, trifloxystrobin) [77–107%]; Others (2,4-D, cyromazine, ethirimol, fipronil, pymetrozine) [35–66%].	LC-ESI$^+$-MS/MS	[6]
10 mL ACN +3 mL hexane (pollen); 10 mL			Neonicotinoids (acetamiprid, clothianidin,	LC-ESI$^+$-MS/MS	[7]

QuEChERS (solvent, salts)	dSPE (solvent)	Sample matrix	Pesticides [recoveries]	Analysis method	Reference
ACN for corn syrup with citrate buffer	50 mg PSA + 50 mg C18 + 150 mg $MgSO_4$	Pollen and high fructose corn syrup Wetted (1:4 dilution)	dinotefuran, flonicamid, imidacloprid, nitenpyram, thiacloprid, thiamethoxam) [>88–110%].		
ACN, 1% CH_3COOH, 6 g $MgSO_4$, 1.5 g NaOAc	0.6 g $MgSO_4$, 0.2 g PSA	Tomato	Azoles (bromuconazole, cyproconazole, difenconazole, diniconazole, epoxiconazole, flutriafol, hexaconazole, imazalil, myclobutanil, penconazole, propiconazole, thiaphanate methyl, triadimefon, triadimenol, triflumizole) [92–106%]; Carbamates (carbaryl, carbofuran, chlorpropham, cycloate, diethofencarb, ethiofencarb, fenoxycarb, methomyl, oxamyl, pirimicarb) [85–104%]; OPs (azinphos methyl, chlorpyrifos ethyl, chlorpyrifos methyl, diazinon, dimethoate, ethoprophos, fenthion, malathion, monocrotophos, omethoate, parathion methyl, pirimiphos methyl, prothiofos, thiometon) [83–109%]; strobilurin fungicides (azoxystrobin, kresoxim methyl, trifloxystrobin) [94–104%]; phenyl or benzoyl ureas (diuron; chlorfluazuron, hexaflumuron, lufenuron) [98–106%]; pyrethroids (bifenthrin, cypermethrin, deltamethrin, fenpropathrin) [93–112%]	LC-ESI$^+$-MS/MS	[8]
ACN, 4 g $MgSO_4$, 1 g NaCl	30 mg PSA, 150 mg $MgSO_4$	Leaf vegetable (pakchoi, rape, crown daisy, amaranth, spinach, lettuce)	Anilide fungicide (metalaxyl) [80–115%]; aryloxyphenoxypropionate herbicide (fluazifop-methyl) [83–119%]. OP (chlorpyrifos) [84–111%]; pyrethroid (Lambda-cyhalothrin) [81–117%].	LC-ESI$^+$-MS/MS GC-ECD	[9]
10 mL ACN, 4 g $MgSO_4$, 1 g NaCl	50 mg PSA, 100 mg $MgSO_4$	Fruits and vegetables (apple, cabbage, carrot, tomato)	Carbamates (aldicarb, baycarb, carbaryl, ethiofencarb, methiocarb); [88–120%]; OPs (azinphos-	LC-ESI$^+$-MS	[10]

QuEChERS (solvent, salts)	dSPE (solvent)	Sample matrix	Pesticides [recoveries]	Analysis method	Reference
			methyl, malathion, methidathion, pirimiphos-methyl [58–71%], etrimfos, pyraclofos, phosalone) [81–120%]; methiocarb-sulfone [72–87%]		
14 mL 1% CH_3COOH in ACN, 6 g $MgSO_4$, 1.4 g NaOAc, 4 g NaCl	All 900 mg $MgSO_4$ and 150 mg PSA >5% fat content also 150 mg C18 <5% colorless to pale extract color, no other sorbents <5% fat content with color (carotenoids/ chlorophyll content high) 45 mg GCB	Food commodities (citric fruits, vegetables, tree nuts, eggs, dairy products, meat, poultry, edible oils, chocolate, coffee, beverages)	OPs (acephate, azinphos-methyl, chlorpyrifos, chlorpyrifos-ethyl, diazinon, dimethoate, disulfoton, demeton-S, demeton-S methyl, ethion, fenamiphos, fenitrothion, fenthion, malathion, methamidophos, methidathion, mevinphos, monocrotophos, omethoate, formothion, parathion, parathion-methyl, phorate, phosalone, phosmet, phosphamidon, propetamphos, terbufos, tetrachlorvinphos, triazophos, trichlorfon, dicrotophos, edifenphos, fosthiazate, isofenphos-methyl, naled, phoxim profenofos, tolclofos-methyl, vamidothion, cadusafos, tribufos, coumaphos, dichlorvos, ethoprophos, isocarbophos, phenoate, quinalphos) PSA [84–107%]; PSA/C18 [83–111%]; PSA/GCB [83–110%] at 10 µg/kg; carbamates (aldicarb, benfuracarb, carbaryl carbofuran, EPTC, fenobucarb, formetanate HCl, isoprocarb, methiocarb, methomyl, molinate, oxamyl, pirimicarb, propamocarb, thiobencarb, thiocarb) PSA [83–106%]; PSA/C18 [85–111%]; PSA/GCB [87–110%] at 10 µg/kg; OP and carbamate degradates (sulfones, sulfoxides) and carbamate degrades (3-hydroxycarbofuran,	LC-ESI$^+$-MS/MS	[11]

QuEChERS (solvent, salts)	dSPE (solvent)	Sample matrix	Pesticides [recoveries]	Analysis method	Reference
			methiocarb) [92–114%] at 10 μg/kg		
(A) 10 mL 1% CH_3COOH in ACN, 4 g $MgSO_4$, 1 g NaCl, Method A, citrate buffer (1 g Na citrate dehydrate, 0.5 g Na_2H citrate sesquihydrate) (B) LLE with 1% HCOOH in acetone	SPE Oasis HLB	Milk (10 mL)	Azoles (azaconazole, epoxiconazole, fenbuconazole, paclobutrazol, thiabendazole, triflumizole) [(A) 82 to >130, (B) 35–114%]; carbamates (aldicarb, carbaryl, carbofuran, diethofencarb, iprovalicarb, methiocarb, methomyl, propamocarb, promecarb, thiophanate-methyl) [(A) <30 to >130%; (B) <30–138%]; neonicotinoids (acetamiprid, imidacloprid, thiacloprid) [(A) 67–123% (B) 83–124%]; benzoyl and phenylureas (diflubenzuron, isoproturon, linuron, metobromuron, metoxuron, monolinuron, pencycuron) [(A) 91 to >130%; (B) <30 to >130%]; sulfonyl ureas (chlorsulfuron, cinosulfuron, iodosulfuron methyl, triasulfuron, thifensulfuron methyl (A) <30–107%; (B) [<30–87%]; triazines (atrazine, metribuzin, propazine, sebuthylazine, simazine, terbuthylazine) [(A) 63 to >130%; (B) <30 to >130%].	LC-ESI$^+$-MS/MS	[12]
15 mL ACN with6.0 g $MgSO_4$ and 1.5 g NaCl	SPE Envicarb (GCB) + SPE Silica (pyrethroids) SPE C18 (pyrethroid degradates)	15 g (A) lettuce, pepper, onion, carrot, broccoli (B) Apple, grape, tomato, orange, banana	Pyrethroids (bifenthrin, cyfluthrin, cyhalothrin, cypermethrin, deltamethrin, esfenvalerate, permethrin) [(A) 49–11%; (B) 50–115%]; pyrethroid metabolites (3-PBA, DCCA, 4-F-3-PBA, DBCA, MPA [(A) 73–136%, (B) 61–121%])	LC-ESI$^+$-MS/MS	[13]
10 mL ACN rinse with 1 mL ACN, citrate buffer (4 g $MgSO_4$, 1 g NaCl, 0.5 g Na_2Hcitrate-	High fat (wheat flour, rolled oats, wheat germ):	Wheat flour and wheat germ shown %	OCs and other halogenated pesticides (aldrin, alachlor, benfluralin, dichlobenil, dieldrin (58–76%),	GC-EI-MS/MS	[14]

QuEChERS (solvent, salts)	dSPE (solvent)	Sample matrix	Pesticides [recoveries]	Analysis method	Reference
$1.5H_2O$, 1 g Na_3citrate dihydrate	salt, PSA, C18 Rich in carotene and chlorophyll (red pepper): salt, PSA, GCB Others (fruits and vegetables) salt, PSA		heptachlor, HCHs, heptachlor epoxide, hexachlorobenzene (42–77%), endosulfan, endosulfan sulfate, iprodione, pendimethalin, trifluralin, triallate, vinclozolin) [most > 80–105% exceptions in brackets]; OPs (bromfenvinphos-methyl, bromophos-methyl, chlormephos, chlorpyrifos, coumaphos, diazinon, dichlorvos, heptenophos, ethoprophos, fenchlorphos, fenthion, fenitrothion, isofenphos, isofenphos-methyl, malathion, mevinphos, parathion, parathion-methyl, tolclofos-methyl) [>80%]; pyrethroids (λ-cyhalothrin, cypermethrin, deltamethrin, esfenvalerate, fenvalerate, flucythrinate, permethrin) [72–103%]		
EtOAc or ACN, 4 g $MgSO_4$, 1 g NaOAc	100 mg PSA, GCB, Zr-Sep+ or C18 or mix of all at 50 mg each	Soya-based nutraceutical–wetted	78–92% of pesticides in 70–120% with ethyl acetate; 3–28% with acetonitrile	GC-EI-MS/MS	[15]
15 mL 1% CH_3COOH in ACN, 6 g $MgSO_4$, 1.5 g NaOAc	200 mg PSA, 600 mg $MgSO_4$	Parsley, lettuce, spinach	Azoles (cyproconazole, difenoconazole, epoxiconazole, penconazole, propiconazole, tebuconazole, triadimefon, triadimenol, triflumizole) [90–100%]; carbamates (carbaryl, carbofuran, carbosulfan, ethiofencarb, fenoxycarb, methiocarb, oxamyl, pirimicarb) [78–111%]; OPs (chlorpyrifos, diazinon, dichlorvos, malathion, dimethoate, profenofos, prothiofos) [86–106%]; pyrethroids (bifenthrin, cypermethrin, deltamethrin, tau-fluvalinate) [98–102%];	LC-ESI$^+$-MS/MS	[16]

QuEChERS (solvent, salts)	dSPE (solvent)	Sample matrix	Pesticides [recoveries]	Analysis method	Reference
			neonicotinoids (acetamiprid, imidacloprid, kresoxim-methyl, thiamethoxam) [77–91%]; strobilurin fungicides (azoxystrobin, trifloxystrobin) [87–103%].		
15 mL EtOAc, 4 g $MgSO_4$, 1.5 g NaCl	Freeze-out, 100 mg Al2O3, 60 mg C18, 600 mg $MgSO_4$	(5 g) bovine liver and muscle	Azoles (tebuconazole, tebufenozide) [73–109%], Benzoylphenylurea (triflumuron) [77–91%]; neonicotinoids (thiacloprid, thiamethoxam) [71–85%]; strobilurin fungicide (trifloxystrobin) [82–94%], other (Spinosyn D) [70–78%]	LC-ESI$^+$-MS/MS and GC-EI-MS	[17]
ACN, 4 g $MgSO_4$, 1g NaCl, 0.6 g Na_2Hcitrate sesquihydrate, 1 g Na_3citrate dihydrate	Freeze-out followed by dSPE with 25 mg PSA and 150 mg $MgSO_4$	Wheat flour (wetted), fruits and vegetables	Organophosphorus pesticides (chlorpyrifos, chlorpyrifos-methyl, fenitrothion, malathion quinalphos) [wheat flour 99–104%]; pyrethroids (bifenthrin, λ-cyhalothrin) [wheat flour 93–99%]; strobilurin fungicides (azoxystrobin, trifloxystrobin) [wheat flour 103–106%] Azoles (difenconazole, tebuconazole) [88–96%]; carbamates (aminocarb, fenobucarb, prochloraz, propamocarb, thiobencarb) [73–108%]; neonicotinoids (acetamiprid, clothianidin, imidacloprid, nitenpyram, thiacloprid, thiamethoxam) [wheat flour 76–102%]; phenylureas (diflubenzuron, flufenoxuron, lufenuron, monolinuron) [wheat flour 86–98%]	GC-EI-MS/MS LC-ESI$^+$-MS/MS	[18]
10 mL ACN, 4 g MgSO4, 1 g NaCl	1 g EMR-Lipid, 1.6 g $MgSO_4$, 1 g NaCl	10 g olive oil or avocado	Azoles (difenoconazole, paclobutrazol, penconazole, tebuconazole, tetraconazole) [76–116%]; carbamates (carbaryl, carbendazim, carbofuran, methomyl) [77–117%]; OPs (acephate, azinphos-	LC-ESI$^+$-MS/MS	[19]

QuEChERS (solvent, salts)	dSPE (solvent)	Sample matrix	Pesticides [recoveries]	Analysis method	Reference
			methyl, chlorfenvinphos, chlorpyrifos [45–51%], chlorpyrifos-methyl [66%], diazinon [102–121%], dimethoate, fenamiphos, fenthion, malathion, methamidophos [60–67%], pirimiphos-methyl, quinalphos, trichlorfon) [71–103%]; neonicotinoids (acetamiprid, imidacloprid, kresoxim-methyl, thiacloprid, thiamethoxam) [82–102%]; phenylureas (chlorotoluron, diuron, flufenoxuron, isoproturon) [73–99%]; strobilurin fungicides (azoxystrobin) [92–96%]		
10 mL ACN with 0.68 mL HCOOH, 2.5 g NaCl	30 mg PSA, 100 mg C18, 60 mg GCB, 150 mg $MgSO_4$	pepper	Neonicotinoid (thiacloprid); spirotetramat and its metabolites [100; 76–89%] at 5 μg/kg	LC-ESI$^+$-MS/MS	[20]
1% CH_3COOH in ACN, 2 g $MgSO_4$ + 500 mg NaOAc	125 mg PSA and 375 mg $MgSO_4$	Bivalve Scrobicularia plana	OCs and related halogenated pesticides (alachlor, aldrin, cyhalofop-butyl, DDD, DDE, DDT, endosulfan, endosulfan sulfate, endrin, HCB, heptachlor, heptachlor epoxide, lindane, mirex, methoxychlor, metoachlor, trifluralin) [81–119%]; OPs (azinphos-methyl, chlorpyrifos, diazinon, dichlorvos, dimethoate, fenamiphos, fenitrothion, fonofos, malathion, methamidophos, parathion, parathion-methyl, phosmet, tetrachlorvinphos) [81–110%] Pyrethroids (cyfluthrin, cyhalothrin, cypermethrin (6%), deltamethrin)[94–114%]; triazines (atrazine, cyanazine, metribuzin, propazine, propyzamide, simazine, terbuthylazine) [85–105%].	GC-EI-MS/MS	[21]
10 mL ACN, 4 g $MgSO_4$, 1 g NaCl	200 mg $MgSO_4$, 200 mg C18	Parsley, basil, mint, thyme, salvia	Carbamates (aldicarb, asulam, benfuracarb, benomyl, benthiocarb,	LC-ESI$^+$-MS/MS	[22]

QuEChERS (solvent, salts)	dSPE (solvent)	Sample matrix	Pesticides [recoveries]	Analysis method	Reference
			carbaryl, carbendazim, diethofencarb, ethiofencarb, fenobucarb, fenoxycarb, isoprocarb, oxamyl, methiocarb, pirimicarb, propamocarb, promecarb, propoxur) [72–98%] at 2 μg/kg		
10 mL ACN	150 mg Z-Sep+ and 150 mg $MgSO_4$	Edible oils (olive, sunflower, maize, linseed and sesame oils) (3:7 dilution with water)	Carbamates (aldicarb, asulam, benomyl, benthiocarb, carbaryl, carbendazim, carbofuran, diethocarb, ethiofencarb, fenobucarb, fenoxycarb, isoprocarb, oxamyl, methomyl, methiocarb, metolcarb, napropamid, pirimicarb, promecarb, propamocarb, propoxur, thiodicarb) [71–104%]	LC-ESI$^+$-MS/MS	[23]
10 mL ACN, followed by freeze-out (−20°C for fat precipitation)	150 mg PSA, 40 mg activated charcoal sorbent, 300 mg $MgSO_4$	Edible oils (rice bran and nut oil)	OCs (aldrin, chlordane, dieldrin, DDD, DDE, DDT, endosulfan, endrin, HCHs, heptachlor) [70–103%]; OPs (dichlorvos, chlorpyrifos, diazinon, fenitrothion, malathion, parathion, parathion methyl, phorate, quinalphos, profenofos, phosmet, phosalone) [67–96%]; Pyrethroids (allethrin, cyfluthrin, cypermethrin, deltamethrin, flumethrin) [68–88%] at 20 ng/g	GC-NCI-MS/MS	[24]
1% CH_3COOH in 10 mL ACN, 4 g $MgSO_4$, 1.7 g NaOAc	40 mg PSA, 150 mg $MgSO_4$	Orange juice	Azoles (bromuconazole, difenoconazole, epoxiconazole, penconazole, propiconazole, tebuconazole, tebufenozide, tetraconazole, thiabendazole) [89–117%]; carbamates (carbaryl, carbofuran, carboxin, mecarbam, thiobencarb) [81–101%]; neonicotinoids (acetamiprid, thiacloprid) [101–106%]; OPs (diazinon, dicrotophos, dimethoate, ethoprophos, fenamiphos,	LC-ESI$^+$-MS/MS	[25]

QuEChERS (solvent, salts)	dSPE (solvent)	Sample matrix	Pesticides [recoveries]	Analysis method	Reference
			monocrotophos, o-methoate, triazophos) [82–113%]; Phenylureas (diuron, linuron, monolinuron) [90–101%]; strobilurin fungicides (azoxystrobin, dimoxystrobin, picoxystrobin) [84–112%].		
1% CH_3COOH in 10 mL ACN, 4 g $MgSO_4$, 1.7 g CH_3COONa	100 mg PSA, 500 mg C18, 600 mg $MgSO_4$ per 4 mL extract	Coconut water and pulp	Azole carbendazim (59% in water), cyproconazole, difenoconazole, thiabendazole, thiophanate-methyl (172% in water) [72–94%]; carbamate (carbofuran) [115 water and 78% pulp]; neonicotinoid (thiamethoxam) [100% water and 96% pulp].	LC-ESI$^+$-MS/MS	[26]
10 mL ACN with 4 g $MgSO_4$ and 1 g NaCl	50 mg PSA, 100 mg C18, 100 mg $MgSO_4$	Meats (high proteins and fats)	Pyraclostrobin, propiconazole, isopyrazam [76–94%] at 5 μg/kg	LC-ESI$^+$-MS/MS	[27]
10 mL ACN with 4 g $MgSO_4$ and 1 g NaCl	25 mg PSA +150 mg $MgSO_4$	Soil (wetting by diluted 1:1 with H_2O)	Neonicotinoids (acetamiprid, clothianidin, imidacloprid, thiacloprid, thiamethoxam) [94–105%]	LC-ESI$^+$-MS/MS	[28]

Table 1. Modified QuEChERS methods for pesticides.

The selection of dSPE sorbent also depends on the target list of pesticides. The use of GCB can reduce recoveries of some pesticides including planar pesticides such as carbendazim, coumaphos, and other pesticides including prochloraz, boscalid, and pyraclostrobin due to strong absorption onto GCB [14]. The use of 25% toluene solution (v/v) can desorb planar pesticides and improve recoveries. The mass of dSPE sorbent is also optimized with reduction of mass improving recoveries for strobilurin fungicides and neonicotinoids along with other problematic pesticides [3]. The original QuEChERS method used 25 mg PSA per mL of extract, but others have increased PSA to 50 mg per mL of acetonitrile extract to obtain recoveries >77% [8].

The original QuEChERS version included no pH control, while current methods use acetate of citrate buffer for pH control to address pesticides that are partially ionized or those that degrade particularly at basic pH conditions such as observed for captan, folpet, dichlofluanid, and tolylfluanid [25, 26]. The buffers are selected as they allow for buffering to pH 4–5.5 for acid sensitive pesticides with minimal loss of base-sensitive pesticides. Some food commodities such as coconut water and pulp also see reduced co-extracts with use of acetate buffer [26]. Comparison of different QuEChERS including the original (salt only), CEN EN 15662 Standard Method (citrate buffer), and AOAC method (acetate buffer) show that recoveries ≥80% can be

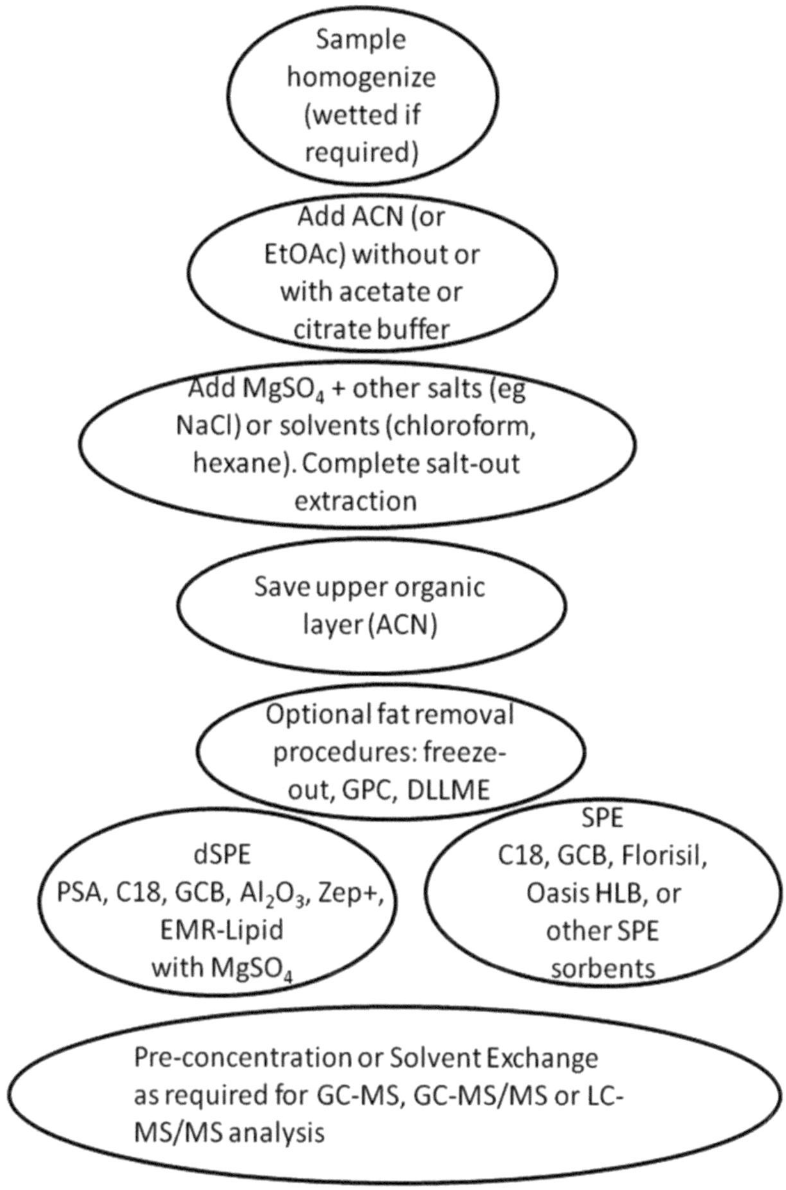

Figure 1. QuEChERS and modified QuEChERS approaches.

obtained for all methods for most pesticides (chemical classes including azoles, carbamates, organophosphorus pesticides, and strobilurin fungicide) in fruit and vegetable matrices analyzed by both GC and LC-MS/MS methods [32]. Acetate buffer pH 4.8 and citrate buffer 5.0–5.5 are used for low pH susceptible compounds such as thiabendazole and imazalil [25]. Low pH samples such as orange juice (pH~3.5) also need pH adjustment during extraction to efficiently extract pesticides of a range of polarities [25]. C18 cleanup decreased the differences in recoveries of the acetate and citrate buffer QuEChERS approaches and was generally found to further improve recoveries [32]. Target analytes with the lowest recoveries included folpet (63–69%) and tolylfluanid (63–71%) analyzed by GC and pymetrozine (31–82%) and tolylfluanid (60–76%) analyzed by LC methods [32]. Ethyl acetate instead of acetonitrile (extraction solvent) has also been used particularly for GC-amenable pesticides [15, 17, 31, 33], but the dSPE is generally more effective with acetonitrile and in some matrices such as peas, the co-extractives may increase significantly when ethyl acetate is used as the extraction solvent [32]. Others have found that the number of GC-amenable pesticides increases with the use of ethyl acetate rather than acetonitrile and good recoveries were obtained with dSPE using a mixture of PSA, GCB, PSA, and Zr-Sep$^+$ [15]. Recoveries improved for cleanup of extracts for analysis of OPs and carbamates by LC-MS/MS (egg products) when acetonitrile, rather than ethyl acetate, was used, and when ethyl acetate was used, recoveries >120% were reported even when followed by dSPE [31]. The use of freezing-out after ethyl acetate salt-out extraction can remove the high lipid content in the co-extracted matrix and if this is followed by C18 and Al_2O_3 addition for removal of lipophilic compounds, fatty acids, sugars, and other acidic compounds (along with $MgSO_4$ to handle water content and high protein content of extracts) it provides better recoveries than when only dSPE with PSA, C18, and Al_2O_3 combinations was used [17]. Buffering of the ethyl acetate extraction can also improve recoveries particularly when the sample matrix is acidic, but care should be taken to minimize ionization of the acidic pesticides (which subsequently increases their solubility in the aqueous phase) [33].

Acetone is a poor extraction solvent and has been found to poorly recover polar analytes such as acephate and cyromazine [31]. Addition of hexane to acetonitrile prior to salt-out has been used to improve recoveries of OPs for bee samples that contain co-extracted beeswax with exception of diazinon and coumaphos that observed a drop in recoveries of 22 and 12%, respectively [5]. Recoveries of neonicotinoids from pollen also improved with addition of hexane to acetonitrile due to the high wax content [7]. Chloroform has also been added to acetonitrile to reduce the amount of acetonitrile remaining in the aqueous phase after phase separations and to further improve the partitioning of polar OPs (methamidophos and acephate) into acetonitrile [29].

For food commodities, the recoveries of analytes analyzed by LC-MS/MS (OPs, azoles, sulfonylureas) increased with dSPE following the salt-out acetonitrile extraction, while for analytes (OCs, OPs, pyrethroids) analyzed by GC-MS/MS, recoveries often decrease into an acceptable range of 70–120% [30]. PSA can bind some analytes strongly such as cinosulfuron that observed 20% decrease in recoveries [30]. Some OPs may exhibit better recoveries with GC-MS/MS rather than LC-MS/MS methods as observed for acephate and methidathion [30]. GC-amenable pesticides tend to include the more lipophilic pesticides, particularly OCs and pyrethroids that have a higher tendency to be extracted with the fatty acid matrix components. If the sample has a low water content, a wetting step is often used; however, if the sample matrix has a high fat content such as wheat flour (5 mg/mL extract) and wheat germ (45 mg/mL) then removing this wetting step (using the Ultra Turrax) will avoid the potential for target analytes such as OCs and pyrethroids to partition into the fatty layer that can form when water is present [14]. QuEChERS method has also been used with a freeze-out step prior to dSPE with PSA for sample matrices with higher levels of co-extracts including lipids (or waxes and sugars) such as wheat flour and citrus extracts [18]. This step can minimize the need for use of other dSPE sorbents. PSA with C18 has improved the recoveries of neonicotinoids from pollen and high fructose corn syrup when the sample is diluted in water (1:4 or 1:8) prior to extraction with acetonitrile (neonicotinoids would be protonated under acidic conditions such that buffers are not used during the salt-out extraction) [7]. Extracts from soil samples also had better recoveries for neonicotinoids when extracted without buffering of acetonitrile (along with salt-out with $MgSO_4$ and NaCl) [28]. C18 (200 mg) alone was used for extract cleanup for analysis of carbamates by LC-MS/MS and found to be better than other dSPE sorbents [22]. The addition of 200 mg of $MgSO_4$ was also used to improve the removal of water so that the evaporation of organic solvent was quicker. QuEChERS with acetate buffer observed low recoveries for PSA + C18 when larger amounts of sorbent were used such that it is often preferred to use only 50 mg C18 [27]. PSA without C18 or GCB was found to provide better recoveries and precision for neonicotinoids in soil [28]. C18 can result in poor recoveries of some more nonpolar GC-amenable analytes (recoveries <70 or >120%) when the sample matrix has a high fat content and, under these situations, Zr-Sep+ has been used to remove lipids [15]. Zr-Sep+ was also used for cleanup of extracts from high fat content edible oil samples reducing matrix effects better than observed with PSA and C18 [23]. Activated charcoal with PSA has also been used for edible oils [24]. A new material called enhanced matrix removal (EMR)-Lipid was also found to perform similar or better than Zr-Sep+ or PSA+C18 for high fat content vegetable matrices with good recoveries for azoles, OPs, neonicotinoids, and phenylureas [19].

The addition of protectants including 3-ethoxy-1,2-propanediol and D-sorbitol prior to GC-MS/MS analysis can also minimize strong interactions of target analytes and matrix with the injector liner or GC column [14, 15, 21]. Re-acidifying extracts after cleanup with acetic or formic acid have also been used to improve peak shapes and response for GC-MS or LC-MS/MS methods and protect analytes that are sensitive to degradation at high pH [18].

QuEChERS approach does not always provide adequate recoveries at low concentrations and issues with large matrix peaks can still be observed in some separations of difficult matrix samples. Consequently, QuEChERS method has been modified to use cartridge SPE cleanup rather than dispersive SPE (**Figure 1**) [12, 13, 34]. Recoveries of pyrethroids and their metabolites improved with the use of cartridge SPE rather than dSPE with 42% of recoveries ≥90%, 70% were ≥80%, 90% were ≥70%, although a range in recoveries was still observed [13]. Metabolites 3-PBA and 4-F-3-PBA did not elute from GCB such that C18 SPE was selected and for some food commodity matrices, a second SPE step with silica or C18 was required [13]. A tandem GCB and PSA cartridge has been used for the cleanup of soil extracts after salt-out acetonitrile extraction for the analysis of range of pesticide classes including azoles, Ops, and pyrethroids [34]. For a wide range of chemical classes of varying polarity, Oasis® HLB (hydrophilic liquid balance) (SPE) was used after the acetonitrile with citrate buffer salt-out extraction to remove additional co-extract matrix components [12]. Although C18 can also provide good recoveries, it is more prone to clogging problems from turbid extracts (in food matrices extract may contain lipids and proteins) such that Oasis HLB is often preferred (**Table 2**) [12].

For some basic analytes, such as pymetrozine which is highly polar, QuEChERS gives poor recoveries as the analyte remains in the aqueous phase as a protonated molecule and adjusting

SPE sorbent (mg)	Elution solvent (volume mL)	Sample type	Pesticide chemical classes [average recoveries %]	Analysis method	Reference
C18 SEP-PAK (500)	DCM (5)	Urine (diluted 1:1 with H2O)	Azoles, OCs, OPs, selected neonicotinoids (kresoxim methyl), pyrethroids [62–109%] Azoles, carbamates, neonicotinoids, phenylureas, strobilurin fungicides [61–101%]	GC-EI-MS/MS LC-ESI$^+$-MS/MS	[36]
C18 Empore extraction disks	ACN (20)	Water	OP (temephos and its degradation products)	LC-ESI$^+$-MS	[37]
C18 (200)	ACN (5.5)	Water	Carbamates [90–99%]	LC-ESI$^+$-MS	[38]
C18, top,+ aminopropyl, bottom	Not specified	Dust (ultrasonic ext with methylene chloride)	Pyrethroids and metabolites [51–101%, resmethrin 23%]	GC-EI-MS/MS	[39]
C18 (500)	MeOH (3)	Urine	Pyrethroid metabolites [90–98%]	GC-EI-MS	[40]
C18 (500)	EtOAc (5)	Air sorbents (filters, polyurethane foam, XAD-2, Tenax-TA), PSE EtOAc	OCs and OPs [80–110%]	GC-NCI-MS	[41]
C18 (500) followed by DLLME	MeOH (1.5)	Water	OCs, OPs, pyrethroids, selected carbamates (carbaryl, pirimicarb) [79–94%]	GC-EI-MS	[42]
ProElut C18 (200)	DCM:MeOH (9:1)	Blood serum	OPs [90–118%] 2.7 ng/mL	GC-EI-MS/MS	[43]
OMICs C18 TIP, µSPE	ACN (0.05)	Wheat (ACN pH 5 ext.)	OPs	LC-ESI$^+$-MS/MS	[44]
Activated carbon µSPE, (100)	EtOAc (2.5)	Vegetables and fruits (microwave ext. with hexane)	OPs [92–105%]	GC-EI-MS	[45]
CleanInert TPT (three materials) (remove pigments, alkaloids, polyphenols)	ACN:toluene 3:1 (20)	teaTea	Carbamates, OCs, OPs, pyrethroids and selected others [88–101%] 5 µg/kg	GC-EI-MS/MS	[46]
GPC + Florisil	Hexane:DCM 5:95 (8)	Milk	OCs	GC-EI-MS	[47]
Sep-Pak C18 (500)	MeOH (10)	Water	Azoles [92–122%], carbamates [OPs [0–108%], strobilurin fungicide [60%], triazine [123–127%] 20 ng/mL	LC-ESI$^+$-MS/MS	[48]
GCB (300)	MeOH (1) + DCM:MeOH 80:20 (5)	Water (pH 2)	Carbamates [83–100%], OPs [78–97%], phenylureas [91–99%], sulfonylureas [90–102%] Protocol 2	LC-ESI$^+$-MS/MS	[49]

SPE sorbent (mg)	Elution solvent (volume mL)	Sample type	Pesticide chemical classes [average recoveries %]	Analysis method	Reference
Oasis HLB (150)	MeOH or ethanol (4)	Tap water (pH 3)	Chlorinated pesticides (alachlor, pentachlorophenol), OP (chlorfenvinphos), triazine (atrazine, simazine), phenylurea (isoproturon) [>80%]	LC-ESI$^+$-MS/MS	[50]
Oasis HLB (150)	MeOH (5), ACN (5)	Water (tap, surface, etc.)	OCs (metolachlor, metazachlor) [76–88%], phenylureas (isoproturon, chlorotoluron, diuron) [86–91%], triazines (atrazine, deethylatrazine, simazine, terbuthylazine) [77–85%]	LC-ESI$^+$-MS/MS	[51]
Oasis HLB (200)	EtOAc (6)	Water	OCs [85–116%], OPs [91–112%], pyrethroids [92–113%], triazines [92–112%]	GC-EI-MS and GC-EI-MS/MS	[52]
Oasis HLB (60)	DCM (1) + MeOH (1)	Water pH 2.5	OCs [55–91%], OPs [35–102%], pyrethroids [74–92%] Azoles [78–91], carbamates [86–90%], strobilurin fungicides [77–92%], phenylureas [88–98%]	GC-EI-MS LC-ESI+-MS/MS	[53]
Envir-carb+NH_2-LC	ACN:toluene 3:1 (25)	Berries (ACN salt-out ext.)	OCs, OPs, selected azoles, and other GC-amenable pesticides	GC-EI-MS	[54]
Oasis HLB (60) or Strata®-X (200)	MeOH (1)	Water (NH_4Ac addition prior to SPE)	Neonicotinoids [85–104%]	LC-ESI+-MS/MS	[55]
Oasis HLB (500)	ACN (5)	ChesnutChestnut, shallot, ginger diluted with water (LLE with ACN)	Neonicotinoids [82–95%] at 0.01 mg/kg	LC-ESI+-MS/MS	[56]
C18 (1000)	MeOH (5)	Atmospheric particles collected on filters	Neonicotinoids and strobilurin fungicides [92–101%]	LC-ESI+-MS/MS	[57]
dSPE: SBA-15-NH_2 (polyphenols removal)	ACN:MeOH 7:3	teaTea	Neonicotinoids [73–85%]	LC-ESI$^+$-MS/MS	[58]
Florisil (500)	MeOH (5)	Honey (1 g diluted 3 mL water:MeOH)	Neonicotinoid (thiamethoxam) +fipronil and degradation products [90–102%]	LC-ESI ($^+$or $^-$) – MS/MS	[59]
Oasis HLB (225)	MeOH (5)	Apple-based infant foods (LLE with ACN)	Carbamates and degradates, azole (thiabendazole) [71–95%]	LC-ESI$^+$-MS/MS	[60]
Oasis HLB (10)	ACN (1)	Rice powder (microwave ext – aqueous extract)	Carbamates (aldicarb, carbaryl, carbofuran, isoprocarb, methomyl, metolcarb, propoxur), phenylurea (diuron) [67–103%] at 10 ng/g	LC-ESI$^+$-MS/MS	[61]

SPE sorbent (mg)	Elution solvent (volume mL)	Sample type	Pesticide chemical classes [average recoveries %]	Analysis method	Reference
Zorbax C18 (500)	MeOH: ACN 1:1 (3)	Water	Carbamates [74–93%]	LC-ESI$^+$-MS	[62]
Graphene (30)	Acetone (5)	Water	Carbamates [55–95%]	LC-ESI$^+$-MS/MS	[63]
Graphene (50)	EtOAc (20)	Apple juice	OPs [94–105%]	LC-ESI$^+$-MS/MS	[64]
C18 (1000)	EtOAc (5)	Air sorbents XAD-2, Tenax-TA, polyurethane foam, PSE EtOAc	OPs, Opoxons, and other OP degradation products [70–100%]	LC-ESI$^+$-MS/MS	[65]
C18 (1000)	0.1% HCOOH in EtOAc-2-Propanol-ACN, 10:55:35, (0.425)	Air sorbents XAD-2, Tenax-TA, polyurethane foam, PSE EtOAc	Azole fungicides [80–108%]	LC-ESI$^+$-MS/MS	[66]
CN-SPE (500)	DCM:MeOH 98:2 v:v	Potato, tomato, orange (LLE)	Carbamate (aldicarb and aldicarb sulfone and sulfoxide) [68–89%]	LC-APCI$^+$-MS	[67]
Oasis HLB (60)	MeOH (3)	wastewaterWastewater	Metabolites of triazines, OPs, pyrethroids	LC-ESI ($^+$ or $^-$)-MS/MS	[68]
Bond Elut SAX + Strata-X	Not specified	Meconium samples from babies	Carbamate (propoxur), OPs, OP metabolites (dialkylphosphates), pyrethroids and metabolites, triazoles	LC-ESI ($^+$ or $^-$)-MS/MS	[69]
Strata X-AW	Parent pesticides EtOAc (5); degradates MeOH:HCOOH 90:10 v/v (3)	Meconium samples from babies	OPs (chlorpyrifos, diazinon, malathion), OP degradates, pyrethroids and degradate, carbamates, phenylurea and metabolite, phenoxyacid herbicide	LC-ESI ($^+$ or $^-$)-MS/MS	[70]
Oasis HLB 96 well plate format (30)	Acetone (0.75)	Urine	Metabolites of OPs [51–92%] and pyrethroids [86–97%]	LC-ESI ($^+$ or $^-$)-MS/MS	[71]
Silica SPE (1000) 1.ISOLUTE ENV+ (200) 2.Bond Elut PPL (200)	MeOH (10) 1. DCM/EtOac 1:1 v:v (6) 2. DCM/EtOAc 1:1 v:v (6)	Urine LLE EtOAc Urine diluted with NH_4Ac buffer (25:10)	Oxy-pyrimidine metabolites of diazinon [LLE +SPE 95:106; SPE only 83–114%]	GC-EI/MS LC-ESI$^+$-MS/MS	[72]
Carbograph (100)	Toluene (8)	Honey	Pyrethroid (tau-fluvalinate), OP (coumaphos), Others: amitraz, fipronil, bromopropylate [99–106%]	GC-PTV-EI-MS	[73]
Oasis HLB (200)	MeOH (8)	wastewaterWastewater	Diazinon, IMP, pharmaceuticals	LC-ESI$^+$-QTOF/MS	[74]
Sep-Pak Plus PS -2, C18 (665) or Oasis HLB (225)	ACN (5), followed by EtOAc (3)	Surface water	OPs, triazines, and selected others [76–99%]	LC-ESI$^+$-MS/MS	[75]

SPE sorbent (mg)	Elution solvent (volume mL)	Sample type	Pesticide chemical classes [average recoveries %]	Analysis method	Reference
Oasis HLB (200)	MeOH (5) followed by EtOAc (5)	Surface water	OCs [45–101%], pyrethroids [45–91%] Azoles [84–133%], carbamates [84–140%], neonicotinoids [104–119%], OPs [68–102], triazines [95–164%]	GC-EI-MS/MS LC-ESI$^+$-MS/MS	[76]
Bond Elut Nexus (polymeric)	MeOH + DCM (1)	Water and wastewater (acidified pH 3)	OCs and OPs [70–120%] some selected OPs and OCs outside of range	GC-EI-MS/MS	[77]

Table 2. Solid phase extraction (SPE) methods for pesticides.

the pH of the extraction leads to problems with recoveries of other acidic or basic analytes. Liquid extraction with acetonitrile (without phase separation using salt) can provide better recoveries than QuEChERS for these analytes, as it does not discriminate basic analytes [35].

3. Solid phase extraction for preconcentration or extract cleanup

Figure 2 illustrates the application of solid phase extraction for different sample matrix types. Solid phase extraction is widely used for the preparation of liquid sample matrices including food beverages, biological fluids, and water samples (drinking water, surface water, ground water). It is also widely used as a cleanup step following prior extraction steps for solid samples such as bee products, air sampling sorbent materials, and soil samples (**Table 2**) [36–77]. For solid sample matrices, popular initial extraction approaches include pressurized liquid extraction (PLE), microwave extraction (MAE), ultrasonic extraction, or liquid-solid extraction [41, 45, 57, 61, 66, 78–96]. Often an organic solvent is selected for the initial extraction of pesticides from the solid materials such that the SPE procedure must be adapted to accommodate the organic content of the sample extract to ensure adequate sorption of target analytes or the sample is diluted with water if feasible prior to SPE.

Table 2 shows common SPE sorbents used along with sample matrix type and target chemical classes of pesticides. SPE sorbents include bonded silica phases such as C18 (or less commonly selected C8); polymeric phases with an aromatic moiety to give stronger retention for more aromatic pesticides through π-interactions; Oasis HLB which is made of a copolymer consisting of divinylbenzene and N-vinylpyrrolidone; carbon based sorbents including graphene for removal of pigments; and NH_2-based sorbents for removal of polar matrix components such as sugars and proteins. N-vinylpyrrolidone acts as a hydrophilic group to give the Oasis HLB sorbent a mixed mode of retention and can improve the retention of more polar pesticides that are weakly retained on C18 sorbents. New generation molecularly imprinted polymers have also been used for cleanup of extracts for analysis of OCs [97]. Both the retention of target analytes and matrix co-extracts must be considered when optimizing an SPE procedure with sample pH and volume during loading, type of SPE sorbent, and extraction solvent and volume optimized.

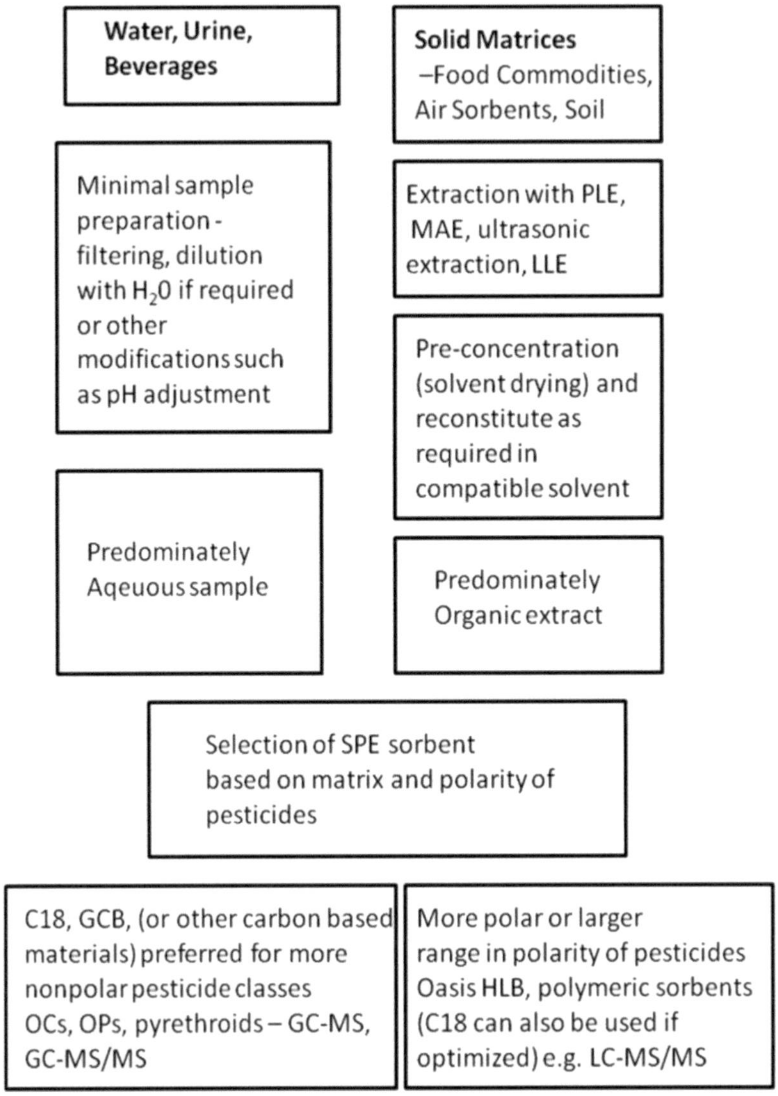

Figure 2. Strategies for extraction utilizing solid phase extract for preconcentration or extract cleanup.

Conditioning solvents for the SPE sorbents are also an important consideration, particularly for liquid extracts that contain an organic solvent from a prior extraction step.

C18 sorbents are more effective at retaining nonpolar pesticides than Oasis HLB with solvent used for extraction often more nonpolar to improve solubility of the target analytes [36]. Selection of elution solvent should also consider a need to dry or evaporate the solvent after SPE either as a preconcentration step or for solvent exchange compatibility for GC-MS or LC-MS/MS analysis. Nonpolar solvents often have higher volatility with ethyl acetate, dichloromethane or mixtures of dichloromethane with methanol commonly selected [36].

For a range of polarity of pesticides (covering both GC- and LC-amenable pesticides), dichloromethane was better at recovering more pesticides (70 and 90 pesticides) compared to methanol (10 and 30 pesticides for LC- and GC-amenable pesticides) [36]. For triazines and phenylureas at acidic sample, pH recoveries were better with Oasis HLB compared to polymeric sorbents (two different nonfunctionalized styrene divinylbenzene (SDVB), hydroxylated SDVB) [51]. Oasis HLB and Strata-X gave good recoveries of neonicotinoids with a lower sample water volume and sorbent amount (60 mg) for Oasis HLB allowing for a small solvent elution volume (1 mL), thereby removing the need for a drying step [55]. Dinotefuran (most polar) and thiacloprid (least polar) had low recoveries due to matrix effects with recoveries improving to 60% with Oasis HLB with a washing step with 5% methanol [55]. For carbamates, better recoveries were observed with Oasis HLB when acetonitrile rather than methanol or ethanol was used as the elution solvent (the lowest recovery observed for methomyl with all solvents) [61]. Oasis HLB, Strata-X, and Strata-C18 were also shown to provide recoveries between 70 and 120% for more pesticides when water samples were acidified to pH 2.5 for both GC- and LC-amenable pesticide classes [53]. Under the optimized method, more pesticides had acceptable recoveries with Oasis HLB as expected from this mixed-mode sorbent [53]; however, recoveries for OCs and OPs varied (**Table 2**), so care should be taken if more nonpolar pesticides are of greatest interest. Under neutral pH conditions, recoveries for OCs and pyrethroids were also more variable when Oasis HLB was used [76]. A larger number of OPs and OCs gave acceptable recoveries with a polymeric sorbent with acidified water samples (Bond Elut Nexus) [77]. Chlorpyrifos and pendimethalin observed low recoveries with Sep-Pak plus PS-2 (C-18) with 5 mL acetonitrile as an elution solvent, but recoveries were improved to >76% with a second elution with 3 mL of ethyl acetate [75]. Carbamates gave good recoveries with C18, while other sorbents including Oasis HLB and carbon-graphitized cartridges gave good recoveries for these carbamates except for pirimicarb and carbofuran [62]. Hydroxylated polystyrene-divinylbenzene copolymer also gave poor recoveries for pirimicarb. Poor recoveries were observed for acephate, chlorpyrifos, and methamidophos in water with C18 SPE, although other OPs observed acceptable recoveries [48]. Graphene is a new SPE sorbent and performs slightly better than C18 or GCB for carbamates except for carbaryl which has lowest recovery of ~55% attributed to stronger π-π interactions with graphene than other sorbents [98]. PRS performed the worst for carbamates of all sorbents tested [98]. Carbon-based sorbents are often selected to remove pigments with the elution solvent selected as toluene or toluene:acetonitrile rather than dichloromethane or ethyl acetate as nonpolar analytes that can bind more strongly to this sorbent material [73, 54]. For extraction of OPs with graphene, ethyl acetate was found to provide better recoveries than dichloromethane or acetonitrile as the elution solvent [99]. Graphene sheets with covalently bonded Fe_3O_4 have also been used for magnetic solid phase extraction (MSPE) of organochlorines in orange juice [100]. Other modified MSPE with Fe_3O_4 including coated carbon nanotubes has been utilized for water or fruit juice extraction of GC-amenable pesticides [99, 101]. Zirconia nanoparticle-decorated calcium alginate hydrogel fibers have been used for extraction of OPs from water and fruit juices [102].

For added selectivity, a molecularly imprinted polymer has been used for SPE sorbent for the analysis of OCs in water, soil, rice, and tea leaves [97]. Micro-SPE has also been used in combination or after extraction methods for recovery of OPs [44–45]. On-line SPE coupled with LC-MS/MS has been used with many of the same sorbents materials described for off-line methods with C18 or C8 and PLRP-s (styrene divinyl-benzene copolymer sorbent) as popular choices [103–106].

4. Other considerations

Solid matrices including soil, sediment, food commodities, and air sampling solid sorbent materials (filter, polyurethane foam, solid sorbents (XAD-2, XAD-4, Tenax-TA)) are extracted prior to an SPE (or dSPE) cleanup step with a variety of approaches including microwave extraction, pressurized liquid extraction (as referred commonly as pressurized solvent extraction), ultrasonic extraction, and traditional solid-liquid extractions [41, 45, 57, 61, 66, 78–96]. These approaches are not selective and the polarity of the organic solvents and choices of additives in these extraction procedures will impact the co-extractive matrix, which necessitate the subsequent SPE or dSPE cleanup choices. In addition, for SPE, aqueous extracts are easier to optimize SPE loading, washing, and elution steps as extracts of organic solvents need careful consideration to ensure adequate retention of target analytes on sorbent materials to prevent washout. The most common solvent choices for pressurized liquid extraction and microwave extraction of solid matrices were acetonitrile, ethyl acetate, acetone, hexane, or combinations of these solvents [81–96]. With microwave extraction, acetone has been added to hexane (2:1) to improve the recoveries for polar OPs, while use of hexane can reduce matrix co-extractives [81]. Ethyl acetate and acetone have been used for microwave extraction of azoles [82]. Reduction in co-extracts has also been reported with acetonitrile rather than methanol or acetone (with microwave extraction) and good recoveries have been reported for OCs and neonicotinoids [83, 84]. Hexane, dichloromethane, ethyl acetate, acetone, and acetonitrile have also been commonly used for pressurized solvent extraction of a large range of polarity of pesticides [41, 57, 65, 66, 85–96].

Acknowledgements

This work was financially supported by the Natural Sciences and Engineering Research Council (NSERC) of Canada.

Abbreviations

ACN	Acetonitrile
APCI	Atmospheric pressure chemical ionization
dSPE	Dispersive solid phase extraction
EI	Electron ionization
EMR-Lipid	Enhanced matrix removal-lipid
ESI	Electrospray ionization

EtOAc	Ethyl acetate
GCB	Graphitized carbon black
GC-MS	Gas chromatography-mass spectrometry
GC-MS/MS	Gas chromatography-tandem mass spectrometry
LC-MS/MS	Liquid chromatography-tandem mass spectrometry
MSPE	Magnetic solid phase extraction
MAE	Microwave assisted extraction
$MgSO_4$	Anhydrous magnesium sulfate
OCs	Organochlorines
Ops	Organophosphorus pesticides
PLE	Pressurized liquid extraction
PSA	Primary secondary amine
QuEChERS	Quick, easy, cheap, effective, rugged, safe
SPE	Solid phase extraction

Author details

Renata Raina-Fulton* and Zhen Xie

*Address all correspondence to: renata.raina@uregina.ca

Department of Chemistry & Biochemistry, Trace Analysis Facility, The University of Regina, Regina, Saskatchewan, Canada

References

[1] Raina-Fulton R, Dunn N, Xie Z. Pesticides and their degradation products including metabolites: Chromatography-Mass Spectrometry, Methods in Mass Spectrometry. Intech, Rijeka, Croatia, 2017

[2] Raina-Fulton R. Journal of AOAC International. 2015;**98**:1163-1170

[3] David A, Botais C, Abdul-Sada A, Goulson D, Hill EM. Analytical and Bioanalytical Chemistry. 2015;**407**:8151-8162

[4] Calatayud-Vernich P, Calatayud F, Simo E, Suarez-Varela MM, Pico Y. Science of the Total Environment. 2016;**541**:33-41

[5] Barganska Z, Slebioba M, Namiesnik J. Molecule. 2014;**19**:2911-2929

[6] Herrera Lopez S, Lozano A, Sosa A, Dolores Hernando M, Fernandez-Alba AR. Chemosphere. 2016;**163**:44-53

[7] Chen M, Collins EM, Tao L, Lu C. Analytical and Bioanalytical Chemistry. 2013;**405**: 9251-9264

[8] Golge O, Kabak B. Food Chemistry. 2015;**176**:319-332

[9] Fan S, Zhang F, Deng K, Yu C, Liu S, Zhao P, Pan C. Journal of Agricultural and Food Chemistry. 2013;**31**:2039–2044

[10] Liu M, Hashi Y, Song Y, Lin J. Journal of Chromatography A. 2005;**1097**:183-187

[11] Chung SWC, Chan BTP. Journal of Chromatography A. 2010;**1217**:4815-4825

[12] Aguilera-Luiz MM, Plaza-Bolanos PP, Romero-Gonzalez R, Martinez Vidal JL, Garrido Frenich A. Analytical Biochemistry. 2011;**399**:2863-2875

[13] Li W, Morgan MK, Graham SE, Starr JM. Talanta. 2016;**151**:42-50

[14] Rasche C, Fournes B, Dirks U, Speer K. Journal of Chromatography A. 2015;**1403**:21-31

[15] Palenikova A, Martinez-Dominguez G, Arrebelo FJ, Romero-Gonzalez R, Hrouzkova S, Frenich AG. Food Chemistry. 2015;**173**:796-807

[16] Esturk O, Yakar Y, Ayhan Z. Journal of Food Science and Technology. 2014;**51**:458-466

[17] Souza R, Pareja L, Cesio MV, Heinzen H. Chromatographia. 2016;**79**:1101-1112

[18] Paya P, Anastassiades M, Mack D, Sigalova I, Tasdelem B, Oliva J, Barba A. Analytical and Bioanalytical Chemistry. 2007;**389**:1697-1714

[19] Lopez-Blanco R, Nortez-Mendez R, Robles-Molina J, Moreno-Gonzalez D, Gilbert-Lopez B, Garcia-Reyes JF, Molina-Diaz A. Journal of Chromatography A. 2016;**1456**:89-104

[20] Li S, Liu X, Dong F, Xu J, Xu H, Hu M, Zheng Y. Food Chemistry. 2016;**192**:893-899

[21] Cruzeiro C, Rodrigues-Oliveira N, Velhote S, Pardal MA, Rocha E, Rocha MJ. Analytical and Bioanalytical Chemistry. 2016;**408**:3681-3698

[22] Natia EA, Moreno-Gonzalez D, Manfo FPT, Gamiz-Gracia L, Garcia-Campana AM. Food Chemistry. 2017;**216**:334-341

[23] Moreno-Gonzalez D, Huertas-Perez JF, Garcia-Campana AM. Talanta. 2014;**128**:299-304

[24] Deme P, Azmeera T, Devi BLAP, Jonnalagadda PR, Prasad RBN, Sarathi UVRV. Food Chemistry. 2014;**142**:144-131

[25] Rizzetti TM, Kemmerich M, Martins ML, Prestes OD, Adaime MB, Zanella R. Food Chemistry. 2016;**196**:25-33

[26] Ferreira JA, Ferreira JMS, Talamini V, Facco JF, Rizzetti TM, Prestes OD, Adaime MB, Zanella R, Bottoli CBG. Food Chemistry. 2016;**213**:616-624

[27] Mu Z, Feng X, Zhang Y, Zhang H. Analytical and Bioanalytical Chemistry. 2016;**408**: 1515-1522

[28] Dankyi E, Gordon C, Carboo D, Fomsgaard IS. Science of the Total Environment. 2014;**499**:276-283

[29] Liu G, Rong L, Guo B, Zhang M, Li S, Wu Q, Chen J, Chen B, Yao S. Journal of Chromatography A. 2011;**1211**:1429-1436

[30] Zainudin BH, Salleh S, Mohamed R, Yap KC, Muhamad H. Food Chemistry. 2015;**172**:585-595

[31] Choi S, Kim S, Shin JY, Kim M, Kim J. Food Chemistry. 2015;**173**:1236-1242

[32] Lehotay SJ, Son KA, Kwon H, Koesukwiwat U, Fu W, Mastovska K, Hoh E, Leepipatpiboon N. Journal of Chromatography A. 2010;**1217**:2548-2560

[33] Jadhav MR, Oulkar DP, Shabeer A, Banerjee K. Journal of Agricultural and Food Chemistry. 2015;**63**:4449-4456

[34] Hayward DG, Wong JW, Shi F, Zhang F, Zhang K, Lee NS, DiBenedetto AL, Hengel MJ. Analytical Chemistry. 2013;**85**:4686-4693

[35] Lacina O, Zachariasova M, Urbanova J, Vaclavikova M, Cajka T, Hajslova J. Journal of Chromatography A. 2012;**1262**:8-18

[36] Cazorla-Reyes R, Fernandez-Moreno JL, Romero-Gonzalez R, Garrido-Frenich A, Martinez Vidal JL. Talanta. 2011;**85**:183-196

[37] Lacorte S, Ehresmann N, Barcelo D. Environmental Science & Technology. 1996;**30**:917-923

[38] El Atrache LL, Sghaier RB, Kefi BB, Haldys V, Dachraoui M, Tortajada J. Talanta. 2013;**117**:392-398

[39] Starr J, Graham S, Stout II D, Andrews K, Nishioka M. Environmental Research 2008;**108**:271-279

[40] Angerer J, Ritter A. Journal of Chromatography B. 1997;**695**:217-226

[41] Raina R, Hall P. Analytical Chemistry Insights. 2008;**3**:111-125

[42] Shamsipur M, Yazdanfar N, Ghambarian M. Food Chemistry. 2016;**204**:289-297

[43] Chang C, Luo J, Chen M, Wu K, dong T, He X, Zhou K, Wang L, Chen D, Zhou Z, Wang X, Xia Y. Analytical Methods. 2016;**8**:4487-4496

[44] Pelle FD, Di Crescenzo MC, Sergi M, Montesano C, Di Ottavio F, Scarpone R, Scortichini G, Compagnone D. Food Additives & Contaminants, Part A. 2016;**33**:291-299

[45] Wang Z, Zhao X, Xu X, Wu L, Su R, Zhao Y, Jiang C, Zhang H, Ma Q, Lu C, Dong D. Analytica Chimica Acta. 2013;**760**:60-68

[46] Zhao C, Ding R, Huo L, Li H, Dong Z, Wang F, Yang G, Lu X, Aboul-Enein HY. Journal of AOAC International. 2014;**97**:1001-1006

[47] Zheng G, Han C, Liu Y, Wang J, Zhu M, Wang C, Shen Y. Journal of Dairy Science. 2014;**97**:6016-6026

[48] Rocha AA, Monteiro SH, Andrade GCRM, Vilca FZ, Tornisielo VL. Journal of the Brazilian Chemical Society. 2015;**26**:2269-2278

[49] Yang XJ, Du Z, Lin A, Yuan Q, Wan P, Wong C. Analytical Methods. 2013;**5**:2083-2092

[50] Barbosa MO, Ribeiro AR, Pereira MFR, Silva AM. Analytical and Bioanalytical. 2016;**408**:8355-8367

[51] Valls-Cantenys C, Scheurer M, Iglesias M, Sacher F, Brauch H, Salvado V. Analytical and Bioanalytical, 2016;**408**:6189-6200

[52] Cruzeiro C, Pardal MA, Rocha E, Rocha MJ. Environmental Monitoring and Assessment. 2015;**187**:669, 21

[53] Donato FF, Martins ML, Munaretto JS, Prestes OD, Adaime MB, Zanella R. Journal of the Brazilian Chemical Society. 2015;**26**:2077-2087

[54] Yang X, Zhang H, Liu Y, Wang J, Zhang YC, Dong AJ, Zhao HT, Sun CH, Cui J. Food Chemistry. 2011;**127**:855-865

[55] Hao C, Morse D, Zhao X, Sui L. Rapid Communications in Mass Spectrometry. 2015;2225-2232

[56] Xie W, Han C, Qian Y, Ding H, Chen X, Xi J. Journal of Chromatography A. 2011;**1218**: 4426-4433

[57] Raina-Fulton R. Journal of Agricultural and Food Chemistry. 2015;**63**:5152-5162

[58] Zhang M, Chen H, Zhu L, Wang C, Ma G, Liu X. Journal of Separation Science. 2016;**39**:910-917

[59] Garcia-Chao M, Agruna MJ, Calvete GF, Sakkas V, Llompart M, Dagnac T. Analytica Chimica Acta. 2010;**672**:107-113

[60] Wang J, Cheung W, Grant D. Journal of Agricultural and Food Chemistry. 2005;**53**:528-537

[61] Song W, Zhang Y, Li G, Chen H, Wang H, Zhao Q, He Q, Zhao C, Ding L. Food Chemistry. 2014;**143**:192-198

[62] Nogueira JMF, Sandra T, Sandra P. Journal of Chromatography A. 2003;**996**:133-140

[63] Shi Z, Hu J, Li Q, Zhang S, Liang Y, Zhang H. Journal of Chromatography A. 2014;**1355**: 219-227

[64] Han Q, Wang Z, Xia J, Zhang X, Wang X, Ding M. Journal of Separation Science. 2014;**37**:99-105

[65] Raina R, Sun L. Journal of Environmental Science and Health, Part B. 2008;**43**:323-332

[66] Raina R, Smith E. Journal of AOAC International. 2012;**95**:1350-1356

[67] Nunes GS, Alonso RM, Ribeiro ML, Barcelo D. Journal of Chromatography A. 2000;**888**: 113-120

[68] Rousis NI, Zuccato E, Castiglioni S. Science of the Total Environment. 2016;**571**: 1349-1357

[69] Meyer-Monath M, Chatellier C, Cabooter D, Rouget F, Morel I, Lestremau F. Talanta. 2015;**138**:231-239

[70] Berton T, Mayhoub F, Chardon K, Duca R, Lestremau F, Bach V, Tack K. Environmental Research. 2014;**132**:311-320

[71] Davis MD, Wade EL, Restrepo PR, Roman-Esteva W, Bravo R, Kuklenyik P, Calafat AM. Journal of Chromatography B. 2013;**929**:18-26

[72] Yokley RA, Shen N, Cheung MW. Journal of AOAC International. 2000;**83**:1229-1238

[73] Notardonato I, Avino P, Cinelli G, Vincenzo-Russo M. Food Analytical Methods. 2016; **9**:1675-1685

[74] Gomez-Ramos M, Perez-Parada A, Garcia-Reyes JF, Fernandez-Alba AR, Aguera A. Journal of Chromatography A. 2011;**1218**:8002-8012

[75] Shinomiya M. International Journal of Environmental Analytical Chemistry. 2013;**93**: 858-871

[76] Charalampous AC, Miliadis GE, Koupparis MA. International Journal of Environmental Analytical Chemistry. 2015;1283-1298

[77] Martins ML, Danato FF, Prestes OD, Adaime MB, Zanella R. Analytical and Bioanalytical Chemistry. 2013;**405**:7697-7709

[78] Mohammadi M, Tavakoli H, Abdollahzadeh Y, Khosravi A, Torkaman R, Mashayekhi A. RSC Advances. 2015;**5**:75174-75181

[79] Ahmadi K, Abdollahzadeh Y, Asadollahzadeh M, Hemmati A, Tavakoli H, Torkaman R. Talanta. 2015;**137**:167-173

[80] Moreno-Gonzalez D, Huertas-Perez J, Garcia-Campana AM, Bosqu-Sendra JM, Gamiz-Gracia L. Journal of Chromatography A. 2013;**1315**:1-7

[81] Merdassa Y, Liu J, Megersa N. Talanta. 2013;**114**:227-234

[82] Wang X, Mao X, Yan A, Tan T, Yang Y, Wan Y. Food Analytical Methods. 2016;**9**:3268-3277

[83] Papadakis E, Kyrgidou A, Vryas Z, Papadopoulou-Mourkidou E. Food Analytical Methods. 2014;**7**:1271-1277

[84] Zeng S, Wu H, Li Z, Wang J, Zhang H, Qian M. Journal of Separation Science. 2015;**38**:121-127

[85] Kadir HA, Abas F, Zakaria O, Ismail IS, Lajis NH. Analytical Methods. 2015;**7**:3141-3147

[86] Lorenzo RA, Racamonde PI, Garcia-Rodriguez D, Carro AM. Analytical and Bioanalytical Chemistry. 2012;**404**:173-181

[87] Masia A, Vasquez K, Campo J, Pico Y. Journal of Chromatography A. 2015;**1378**:19-31

[88] Van Emon JM, Chuang JC. Analytica Chimica Acta. 2012;**745**:38-44

[89] Wang D, Weston DP, Ding Y, Lydy MJ. Archives of Environmental Contamination and Toxicology. 2010;**58**:255-267

[90] Zhang Y, Yang J, Shi R, Su Q, Gao Y, Zhu X. Chromatographia. 2011;**73**:385-391

[91] Wu G, Bao X, Zhao S, Wu J, Han A, Ye Q. Food Chemistry. 2011;**126**:646-654

[92] Homazava N, Aquillon CG, Vermeirssen E, Werner I. International Journal of Environmental Analytical Chemistry. 2014;**94**:1085-1099

[93] Kock-Schulmeyer M, Olmos M, de Alda ML, Barcelo D. Journal of Chromatography A. 2012;**1305**:176-187

[94] Leyva-Morales JB, Valdez-Torres JB, Bastidas-Bastidas PJ, Betancourt-Lozano M. Journal of Chromatographic Science. 2015;**53**:1623-1630

[95] Celeiro M, Llompart M, Lamas JP, Lores M, Garcia-Jares C, Dagnac T. Journal of Chromatography A. 2014;**1343**:18-25

[96] Chitescu CL, Oosterink E, de Jong J, Stolker AAM. Talanta. 2012;**88**:653-662

[97] Gao X, Pan M, Fang G, Jing W, He S, Wang S. Analytical Methods. 2013;**5**:6128-6134

[98] Shi Z, Hu J, Zhang S, Liang Y, Zhang H. Journal of Chromatography A. 2014;**1355**:219-227

[99] Wanibrahim W, Nodeh HD, Aboul-Enein HY, Sanagi MM. Critical Reviews in Analytical Chemistry. 2015;**45**:270-287

[100] Sun T, Yang J, Li L, Wang X, Li X, Jin Y. Chromatographia. 2016;**79**:345-353

[101] Chen F, Song Z, Nie J, Yu G, Li Z, Lee M. RSC Advances. 2016;**6**:81877-81885

[102] Zare M, Ramezani Z, Rahbar N. Journal of Chromatography A. 2016;**1473**:28-37

[103] Lucci P, Nunez O. Journal of Separation Science. 2014;**37**:2929-2939

[104] Liao H, Hsieh C, Chiang S Lin M, Chen P, Wu K. Journal of Chromatography B. 2011;**879**:1961-1966

[105] Sancho JV, Pozo OJ, Hernandez F. Rapid Communications in Mass Spectrometry. 2000;**14**:1485-1490

[106] Mann O, Pock E, Wruss K, Wruss W, Krska R. International Journal of Environmental Analytical Chemistry. 2016;**96**:353-372